Safety at Height

Everybody climbs up or down for some reason or other. Many fall and get injured or, worse, end up killed all over the world, especially in their place of work. Why does it happen? Does it have to be so? Can it be managed better and, if so, how? This book addresses these questions in layman's language, yet with sufficient technical detail to satisfy the more curious and challenge the more ambitious.

In *Safety at Height: A Holistic View of Fall Management*, veteran author Natarajan Krishnamurthy shares his long research and consultancy experience on this subject to offer an overview of falls, methods to manage them, and practical techniques to ensure better safety. This book argues that deaths and major injuries from fall accidents can be prevented by stakeholders knowing more and following guidelines. It looks at the mechanics of falls, accidents in the workplace, and safeguards that can be put in place. Featuring exercises at the end of chapters to underpin learning, this title concludes with unusual fall situations. Through its pages, the reader will develop a good understanding of how to prevent falls across a variety of different real-life scenarios.

This handy guide will be an ideal read for students, researchers, and professionals in occupational safety and health, human factors, and activities where slips, trips, and falls tend to occur.

Natarajan Krishnamurthy ('Prof Krishna') is an international consultant in safety, structures, and computer applications. Now based in Singapore, he has had significant roles in teaching and training, research and consultancy, in India, the USA, and Singapore, underpinning the breadth and depth of his experience. He has authored more than 100 papers and 18 books in his technical specialties and general fields. He has three inventions patented in Singapore.

He is Founder President of the Singapore Society of Steel Structures, Singapore Section of the American Society of Civil Engineers, and the Centre for Workplace Safety and Health at the National Institute of Engineering in Mysuru, India.

Safety at Height
A Holistic View of Fall Management

Natarajan Krishnamurthy

CRC Press
Taylor & Francis Group
Boca Raton London New York

CRC Press is an imprint of the
Taylor & Francis Group, an **informa** business

Cover Design: Natarajan Krishnamurthy. Picture credit: Remix by Natarajan Krishnamurthy from a royalty-free black and white photograph by Lewis Hines, with Lunapic Art (StorytIme 75%).

First edition published 2024
by CRC Press
2385 NW Executive Center Drive, Suite 320, Boca Raton FL 33431

and by CRC Press
4 Park Square, Milton Park, Abingdon, Oxon, OX14 4RN

CRC Press is an imprint of Taylor & Francis Group, LLC

Library of Congress Cataloging-in-Publication Data
Names: Krishnamurthy, Natarajan, author.
Title: Safety at height : a holistic view of fall management / Natarajan Krishnamurthy.
Description: New York, NY : Routledge, 2024. | Includes bibliographical references and index.
Identifiers: LCCN 2023053502 (print) | LCCN 2023053503 (ebook) |
    ISBN 9781032616971 (hardback) | ISBN 9781032648125 (paperback) |
    ISBN 9781032648132 (ebook)
Subjects: MESH: Accidental Falls—prevention & control | Accident Prevention—methods |
    Accidents, Occupational—prevention & control | Accidents, Home—prevention &
    control | Holistic Health
Classification: LCC HV675 (print) | LCC HV675 (ebook) | NLM WA 288 |
    DDC 613.6—dc23/eng/20240228
LC record available at https://lccn.loc.gov/2023053502
LC ebook record available at https://lccn.loc.gov/2023053503

ISBN: 978-1-032-61697-1 (hbk)
ISBN: 978-1-032-64812-5 (pbk)
ISBN: 978-1-032-64813-2 (ebk)

DOI: 10.1201/9781032648132

Dedicated –

To those who love heights, to help them be safe or get safer;

To those who fear heights, to lead them to safe management;
and

To those who don't care, to woo them toward greater heights.

Contents

Preface

Any human activity at height has fascinated me throughout my adult life for three reasons.

-1-

As an engineering student in India in the 1950s, while on a technical education tour to an iron-ore mining facility, my group was trudging wearily up the hill to the control station of the ropeway for the ore buckets transporting the mined ore from the top of the hill to the bottom, arguing how it would be like to ride to the top in one of the empty buckets. Before I knew it, I had accepted the dare to show it could be done!

I climbed the nearly 15 m (50 ft) high tower, lowered myself into one of the slow-moving buckets, squatted within it, and started enjoying the panoramic view around me, trading jokes with my foot-sore classmates.

At high noon, the ropeway stopped for the workers' lunch break, with my bucket hanging midway between two towers!

When guards on their rounds saw me high up in the air, they got into a frenzy and mobilised my rescue. I discovered the reason for their panic as soon as they shouted that I must hold tight to the bucket sides, as otherwise, if I accidentally tripped the latch (yes, I saw it right then!) that held the bucket upright and automatically flipped it upside down to empty it when it reached the bottom, I would have fallen to my death, literally kicking the bucket!

Well, after my slow journey to the top, a tongue-lashing and a formal complaint there, all of us returned home. I was given a verbal thrashing by the college principal and my family elders. The only reason I was not dismissed from college was that I was a top scorer.

I have frequently wondered how I never felt any fear or anxiety while doing it. Certainly, it was because I had not known about the latch that would have meant my death by one false move. Also, climbing trees and boarding running buses were part of every normal youngster's repertoire in my time. But basically, I seemed to have no fear of heights.

Now that I have spent decades learning and teaching risk management, I have realised how stupid of me it was to have accepted the challenge without considering the risk of falling from that height. I shudder every time I think how close I had come to killing myself and ruining the futures of my dependants.

I have been so embarrassed, even ashamed, about this escapade that I have avoided mentioning it or responding to queries about it.

But I confess it now, to explain my inborn fascination with heights, to share my learning first-hand the dangers of moving to and at heights, and to warn wannabe superheroes to think twice before they take any such risk, about whether they or their loved ones, or their employers, can afford the consequences if something goes wrong!

-2-

The second reason is more professional.

In the USA in the 1970s and 1980s, when I worked on structural failure investigations, some of the actions involved climbing heights on ladders and scaffolds. I would follow the lead of my American colleagues in the climbing, as of course they were more experienced in that work.

Even then, I did not think twice about what the risks were, because at the back of my mind was the faith that if my colleagues could do it, it should be safe enough for me to do it too!

But I gradually saw the logic of the safety measures my companions used during these inspections, such as the 'three-point' rule for climbing ladders. The Occupational Safety and Health Administration (OSHA) too had just started prescribing safety rules, and so my learning curve was fast and smooth.

Moreover, the years I spent in the USA also demonstrated to me how much Westerners loved rock climbing and other sports involving heights, and how seriously they planned and trained for them.

Those were my apprenticing years at workplace heights.

-3-

The third reason is the clincher, you may say, my main motivation!

When I got into workplace safety and risk management in Singapore more than two decades ago, mostly focusing on the construction industry, I had the opportunity to climb scaffolds and formwork fully conforming to safety regulations and to offer training courses for supervisors and other safety personnel on the hazards of working at height and their control.

In due course, I was invited to investigate accidents of falling from height and I also carried out personal study and sponsored research into working safely at height.

Soon, the problems of working at height and the many errors of omission and commission that led to accidents, the many human factors contributing to the root causes for falls, and the feasible solutions to these problems fascinated me to the extent that the topic became a passion, a crusade, and a mission for the rest of my life.

***

These three experiences have goaded me into studying more and sharing more of what I learnt from and with fellow engineers. I have conducted hundreds of short courses, made numerous seminar presentations, and published many papers on the subject since 2000 in Singapore, the USA, and India.

I have also patented an invention of a wheelchair modification that would eliminate the falling risk of old or sick or physically challenged users if they chose to get up and walk on their own.

This book is thus a logical outcome of my intent to share my knowledge and opinions on this vital subject.

All over the world, people keep falling all the time from heights, and even at level, for various reasons; many even die or get seriously injured from these falls.

While some of these mishaps may be attributed to unavoidable causes like slipping unexpectedly on a wet surface or ignorantly traversing an unguarded walkway, most of the 'accidents' could have been avoided by proactive planning and proper follow through.

In the following pages, I plan to discuss what happens when someone falls from height or at level and present my views on how to control or manage them. *A la*

Newton, I certainly stand on the shoulders of giants in offering such information and comment.

I shall refrain from the following:

- Invoke Government or Industry Regulations or Codes of Practice as my recommendation;
- Refer to any specific group or organisation as fully endorsed by me or being contrary to my thinking; or
- Include data or mention specific findings from any of my sponsored research not in the public domain, to support my recommendations.

I do this not only to keep this book above controversy or even any appearance of bias, but also because I intend to approach the entire problem of working safely at height holistically, in the sense of addressing the overall and entire ('whole') situation, rather than proffer specific recommendations based on rules.

My aim will be fulfilled if the book is useful to the reader in understanding the complicated subject and putting the knowledge gained to some practical use, or at least if the casual reader enjoys the general and descriptive portions if not the technical and the prescriptive!

N. Krishnamurthy
January 2024

At the very last minute, just before the book went into press, I decided to take the bold step of going public with the only proof extant of my fateful escapade, on Monday, 17 December 1951 (more than seventy-three years ago!), namely the photograph taken around 11 am, by my close friend and classmate Mr. B.V. Sreenivasa Murthy, of me sitting in the bucket, calmly waving to him, while the other four in the group nervously stand by, worried about their part in provoking my rash and irresponsible act. Little did I know then that if I had jiggled around a little more in my perch, I would have fallen, to my death!

PLEASE DON'T TRY THIS!

Acknowledgements

I acknowledge with profound gratitude the support of the following individuals and organisations in the writing and production of my book:

- Mr. Mike Kingdon for permission to use his bungee jump photo, Figure 1.1(a).
- Rent-a-monkey Tree Service, Sandy Utah, USA, for permission to use their tree-cutting picture, Figure 1.4(b).
- White Wolf Pack group of USA, for permission to use the picture of the Mohican steel worker, Figure 1.5(a).
- AskAladdin, the Travel Experts, UAE and USA, for permission to use their pyramid construction picture, Figure 2.3(a).
- The Collector, Montreal, Canada, for permission to use their picture of Roman construction practices, Figure 2.3(g).
- Golden Gate Bridge, Highway & Transportation District, San Rafael, CA, USA, for permission to use material from their websites, leading to Figure 2.4(c) and (e).
- Workplace Safety and Health Council, Ministry of Manpower, Singapore, for permission to use material from their Code of Practice of Working Safely at Heights, Second Revision 2013, leading to Figures 3.5, 3.6(c), 3.7, 4.2(b), and 4.4(c).
- Circus Arts, Byron Bay, NSW, Australia, for permission to use their trapeze net image, Figure 6.1(d).
- Karam Safety Pvt. Ltd., Noida, India, for permission to use material from their website, Figure 10.3(b).
- Prof. Salman Azhar, Auburn University, Alabama, USA, for permission to use the BIM image from his co-authored paper, Ref. 10.5, Figure 10.4(b).
- OSHA of USA, HSE of UK, MDPI of Switzerland, and Wikipedia Commons, for making available in public domain many pictures printed in this book.
- Unsplash, Freepik, Picryl, and many other websites offering photos and line art, for making available copyright-free graphics, some of the fee-free ones of which are used in this book, with citation where required.
- CRC Press, Routledge – Taylor & Francis Group, LLC, UK, for accepting to publish my book in their Focus series, and producing such an attractive work.
- Mr. James Hobbs, Editor II, Ergonomics/Human Factors and Occupational Health and Safety. CRC Press – Taylor & Francis Group, LLC, UK, for initial review, vetting and processing of the book concept and contents, up to acceptance stage.
- Ms. Kirsty Hardwick, Editorial Assistant, CRC Press – Taylor & Francis Group, LLC, UK, for clarifications, continuous guidance, and processing of the various components of the book as needed.
- Ms. Vaishnavi Madhavan, Project Manager, Apex CoVantage India, and her team, for reviewing and finalising my manuscript for printing.
- My clients and research sponsors for helping add to my store of knowledge in this area.

N. Krishnamurthy

1 Introduction

1.1 OVERVIEW OF WORK AT HEIGHT

Working at height has been the most hazardous activity in human history since caveman days and continues to be so even now, in fact increasingly so because we are now building taller and taller structures all the time.

1.1.1 WHY FALLS ARE PARTICULARLY DANGEROUS

Falls have been and continue to be the most common cause of unintentional, that is accidental, injury uniformly all over the world.

Why should this be so?

The one-word answer is Gravity: universal, perpetual, inexorable, and omnipresent force (technically an 'acceleration', which translates to a force) on our planet earth, which affects anything and everything on or close to the earth. The only way to escape it is to become an astronaut and shoot off into space about 400 km above the earth and keep flying at 28,000 km/h, like the International Space Station is doing!

Even airplanes do not fall only because their propellers create an upward thrust on their wings. Helicopters and drones hover aloft only by the downward thrust of their rotors against gravity. Balloons stay up with lighter-than-air gases or hot air.

In all other cases, what goes up must fall down, and when it hits the ground or other base, the object will experience an impact force, which may damage it and (if living) may injure it badly or even kill it.

Working at depth (as in mining or tunnelling) is the reverse of working at height in that in the former, persons who go up must come down, and in the latter, persons must make their way down to reach their goal and also must come back up.

A major difference is that most of the accidents in underground work occur due to rock (or earth) falls and roof collapses, rather than from the risks of working at a different level. The relatively smaller number of falls while climbing down to or up from a depth can be handled in the same way as above ground work – easier possibly because side support is generally available. So, we will not focus on underground work falls separately.

1.1.2 WHY FALL REGULATIONS ARE SO DIVERSE

Why does the entire world NOT have a single regulation for working at height, as gravity is the same all over the world?

This is because:

- The demographics, namely height, weight, and other individual characteristics, vary from region to region, from ethnic group to ethnic group;

DOI: 10.1201/9781032648132-1

- Natural and financial resources vary from country to country;
- Different industries and tasks pose different types and levels of risk, demanding different safeguards;
- Priorities and preferences depend on cultural orientation;
- Investments in fall management depend on the country's GDP and the party in power.

Ultimately, any and all local, regional, and national regulations and codes must and will prevail. Any code clause I mention specifically will be only to mark trends and guide with typical values, but may not be my recommendation!

1.2   WHY PEOPLE GO TO HEIGHTS AND HOW THEY FALL

1.2.1   WHY PEOPLE GO TO HEIGHTS

(a)   To test and prove courage and skill

For thousands of years, youth of the Vanuatu island near Australia have been jumping off a wooden tower with a bungee (vine) cord round their ankles, to prove their courage and claim their brides. Figure 1.1(a) depicts a scene from this rite of passage.

Week-end rock-climbing, mountaineering, and skiing [see Figure 1.1(b),(c),(d)] are personal challenges faced and overcome. Most of the gymnastics, pole vaulting [see Figure 1.1(e)], and other competitive sports are a test of courage combined with skill. All these involve high fall risks.

(b)   To play and have fun

Games and sports have always had their share of jumping and falling. But in contrast to the high pressure of competitive genre discussed previously, people enjoy playing just for the fun of it.

Currently, vacation spots offer – for a fee – exciting events involving heights where participants can experience the thrills that their more talented and better trained peers go through, without the bad consequences that the latter suffer.

For instance, a vacationer can now go paragliding, in tandem with a trained professional; jump from a high platform on to an airbag and not even be shook up; and do a bungee jump of 100 m with no fear.

In this category, there must also be included visitors to unusual tourist spots, who want to retrace the path of ancient civilisations with all their inherent fall dangers, such as narrow, steep steps with no handrails in old forts.

(c)   To meet need, to survive

Most of us have to climb up or down because we have to fetch, check, or fix something; to meet or assist somebody; or to escape from some threat. Examples range from simply changing a bulb in the kitchen to jumping out from a building on fire.

There have been – and still are – situations where risky access to heights and depths is casually negotiated daily because that is the only way people can get some critical need, like water for the home, or earn a livelihood.

FIGURE 1.1 Work at height activities.

(a) Bungee jumping in Vanuatu tribe. [Source and Author: Photo by Mike Kingdon. Link: www.takingthemike.com, with permission from the author.]

(b) Rock climbing [Source: Transferred from Wikipedia to Commons by Zeimusu using CommonsHelper. Author: 2005biggar at English Wikipedia. Link: https://commons.wikimedia. org/wiki/File:Kate-at-fleshmarket.JPG]

(c) Mountaineering [Source: Own work. Author: Benh Lieu Song. Link: https://commons. wikimedia.org/wiki/File:Alpinistes_Aiguille_du_Midi_03.JPG]

(d) Skiing [Source: Own work. Author: Martin Rulsch, Wikimedia Commons. Link: https:// commons.wikimedia.org/wiki/File:2020-01-21_Freestyle_skiing_%E2%80%93_ Snowboarding_at_the_2020_Winter_Youth_Olympics_%E2%80%93_Team_Ski-Snowboard_Cross_%E2%80%93_Finals_%E2%80%93_Small_Final_%28Martin_ Rulsch%29_16.jpg]

(e) Pole vaulting [Source: Archives of Pearson Scott Foresman, donated to the Wikimedia Foundation. Author: Pearson Scott Foresman. Link: https://commons.wikimedia.org/wiki/ File:Pole_Vault_%28PSF%29.png]

FIGURE 1.2   Need to access heights or depths.

(a) Centuries-old stepwell at Abhaneri, Rajasthan, North India [Source: Own work. Author: Chetan. Link: https://commons.wikimedia.org/wiki/File:Chand_Baori_%28Step-well%29_at_Abhaneri.JPG]

(b) Construction worker [Source: Construction Worker on High-Rise – Dhaka – Bangladesh. Author: Adam Jones from Kelowna, BC, Canada. Link: https://commons.wikimedia.org/wiki/File:Construction_Worker_on_High-Rise_-_Dhaka_-_Bangladesh_(12850641585).jpg]

Figure 1.2(a) shows a centuries-old well in North India, which once held excellent pure water is now reduced to sludge due to disuse and lowering of the water table and consequent neglect. The steps are narrow and high and have no guardrails – the rails in the photo are modern additions for tourist safety. Some similar wells are still in use in villages in India.

Figure 1.2(b) shows a construction worker in a third-world country readying the work for casting in a high-rise building. He does not have any safeguard against falls. He does not wear any protective gear and is not attached to any anchor. But that is the only way he can keep the job!

In both cases, the persons involved did not really worry, or even think, about the dangers.

There are many places on earth where school-going children have to climb cliffs without a regular path and cross ravines or streams on a rope strung across, where men reach beyond a ledge high in the air to ready a platform for others to work from, and where women carry heavy headloads over recently cast concrete steps without a guardrail.

Ref. **1.1** gives descriptions and pictures on some of the most dangerous journeys to school in the world.

(d)   In normal work

Even today, many tasks in our normal daily life require somebody to climb heights. Picking coconuts from palm trees is still mostly a human activity, as are servicing of electrical connections and cleaning gutters under sloping roofs.

Even changing a light bulb, which most of us can do casually, involves risks of falling from height – and companies have paid millions of dollars compensation for workers' deaths while changing bulbs!

FIGURE 1.3 Climbing for work.

(a) Steel workers [Source: flickr. Author: Photo taken by flickr Paul Keheler. Link: https://commons.wikimedia.org/wiki/File:Construction_Workers.jpg]

(b) Electrical workers. [Source: Own work. Author: Kunlemessi. Link: www.google.com/search?q=elctrical+workers+wiki+commons&tbm=isch&ved=2ahUKEwieiabKodn_AhV9j2MGHT3iCDUQ2-cCegQIABAA&oq=elctrical+workers+wiki+commons&gs_lcp=CgNpbWcQA1DyJVjTgQFgkYkBaABwAHgAgAE0iAHkA5IBAjEy-mAEAoAEBqgELZ3dzLXdpei1pbWfAAQE&sclient=img&ei=fH6VZJ7bJ_2ejuMPvcSjqA-M&bih=636&biw=1163&rlz=1C1CHBD_enSG726SG726#imgrc=rnE7dbGOKUoMwM]

(c) Roofing workers [Source: LeBlanc Construction Photos 2012. Author: National Institute for Occupational Safety and Health (NIOSH) from USA. Link: https://commons.wikimedia.org/wiki/File:Roofing_workers_fall_prevention_%289253637735%29.jpg]

Certain industries are rife with hazardous tasks. All over the world, regardless of the country's size, GDP, safety culture, or other measures of sociological stability or technological advancement, construction tops the list in workplace fall fatalities, with manufacturing coming a close second.

Many of the tasks involve heights, as depicted in Figure 1.3, and consequently, many accidents and injuries from them are due to falling from height.

The biggest single factor that has made people seek heights is that men find it more convenient and efficient to congregate in high-rise buildings. To build and maintain these tall structures the developers and owners must keep all their personnel safe from fall accidents.

(e) To pray

Most temples and sacred sites around the world are situated on top of hills and mountains, symbolising the devotees' reach towards the heavens. Climbing heights to pray is universal.

Even on level ground, the spires of temples, mosques, and churches stand tall to inspire and beckon to the faithful. Persons in charge of the structures have to access heights for construction, maintenance, and modification, or call to prayer, etc.

The pyramids in Egypt were raised to entomb royalty upon their death, so their souls may rise to afterlife in heaven. Moses is said to have climbed Mount Sinai to receive the Ten Commandments from God.

The myth of the Tower of Babel articulates the belief that one may get closer to heaven by reaching high. Similar myths exist in other religions, of prophets preaching from heights.

Thus, heights are associated with nearness to the Almighty.

(f) To explore the unknown

Some folks will always push the envelope to go where no others have been, to do what no others have done, and to jump higher than others have.

Edmund Hillary's (and his Tibetan companion Tenzing Norgay, to whom it was just another day's work!) scaling Mount Everest in 1953 was one such feat.

An earlier climber (who perished in his attempt) is quoted to have said he was doing it because 'it is there'.

Olympic champions keep struggling to beat their own records, exploring how much higher they can jump.

To these may be added, suicidal as it may seem, and not at all recommended for responsible folks, the recent extreme sport 'parkour', in which practitioners attempt to get from one point to another at a different height and across many gaps and obstacles, in the quickest and most daring way possible, without assisting equipment and without prior planning or rehearsal.

1.2.2 WHY PEOPLE FALL

Obviously, not all people who go to or do something at heights fall and get injured. A truism – not yet realised – is that all falls can be prevented if sufficient prior planning is done and all safeguards are fully and carefully implemented.

It is true that many people continue to manage fall situations with confidence or simply and spontaneously carry out the tasks at height without even being aware of the danger.

At the same time, glaring statistics alert us to the fact that injuries and fatalities due to falls are quite common and large in number all over the world even currently.

Apart from falling from a higher level to a lower one, people fall walking or doing some other activity standing, on a level surface, which in workplace parlance is called 'slips and trips', respectively meaning sliding on slippery surfaces or tripping on obstacles.

Due to a peculiarity of our blood circulation system, people may fall even if they stand still or are suspended from a harness, for a long time. (We will see why, later.)

The simple fact is that people fall because they are unable to – or do not – take care or adopt proper steps to negotiate the difference in level or even to stay or traverse some distance at level, and gravity drags them down to crash with increasing velocity.

Among specific reasons for falls may be cited the following:

(a) Ignorance of the mechanics of falls and their consequences;
(b) Lack of provision of safeguards against falling in the habitat and the workplace;
(c) Negligence of the precautions, guidelines, rules, and safeguards provided for negotiating and working at height;
(d) Abnormal health conditions such as vertigo, advancing age, medication effects, etc.

(Although many seek heights to jump from to commit suicide – the Golden Gate Bridge in San Francisco, USA, being a favourite venue – we will not include such voluntary self-destruction as a topic for our discussion.)

Thus, going to or doing anything at height need not be automatically classified as a threat to health and life. Many factors contribute to the type of danger and the severity of injury, and most of them can be managed better by proper understanding and with appropriate safety measures.

1.3   FACTORS AFFECTING FALL MANAGEMENT

If people have been managing fairly well not to fall from the beginning of time, why worry about it now?

This is simply because modern 'civilised' man has other preoccupations than learning and coping with the inexorable effects of gravity – such as carrying out his duties and make-work tasks and establishing individual or collective superiority over others seeking the same ends to provide a 'better quality of life' or even basically to survive amidst controlled chaos.

Also, because nations still vie for the highest bridge, the tallest building, and the like, the trend towards high-rise construction going higher will not abate in the foreseeable future.

That is why we must continue to battle the falling problem!

The following subsections cover some of the factors that influence the way we view and address fall dangers today.

1.3.1   NECESSITY AND FAMILIARITY

As described in the previous section, our ancestors faced fall dangers that many cannot handle today, without any concern for (or even awareness of) preventive and protective measures.

Even today, some folks manage heights better than others.

How is this possible?

The answer to this is that earlier, right from childhood everybody grew up with the risks of heights (as of other dangers) and so it was imperative for them to avoid the danger. When they could not, they took precautions against falling and getting hurt.

Successive generations learnt from their forbears. Thus, they kept safe automatically, instinctively, like we now use zebra crossings even while glued to our cell phones without paying much attention.

Even in the workplace, there was no fall protection, and employers took no responsibility for falls. When somebody thought of it, the person at risk would wear a rope around his waist and tie it to something at the higher point, so that if he fell, he would not hit the base.

Each worker believed that he alone was responsible for his safety and, hence, watched out for himself to do everything right.

Modern man, particularly growing up in an urban setting, has lost this self-preservative motivation and action, that is 'survival skills'.

Authorities require managements to protect the public and the workforce, according to safety regulations, with safety devices recommended and approved by them. So, the incentive for self-preservation has vanished.

1.3.2   RESOURCES AND ECONOMICS

Where we live and the resources a country or company influence how we manage fall risks. Countries with low GDP do have more accidents at the workplace.

An article on safety recommendations for Bangladesh [Ref. **1.2**] argues that different countries cannot use and may not even need global standards of safety.

As a case in point, today, a villager in India or Philippines can climb a coconut palm 20 m high with nothing but a rope around his feet and simply hugging the tree, as in Figure 1.4(a) – or at best holding on to a rope around the tree.

His counterpart in the USA [see Figure 1.4(b)] is dressed more stylishly (and of course much more expensively!) with a waist belt around the tree and sharp spikes on his shoes to dig into the wood (or foot prongs driven into the tree) to support him as he climbs, works, and descends.

Both work at height and both survive equally well, demonstrating how regional practice can play a big part in safety.

The differences are in how they meet the risk with the resources they have. The interesting fact is that fall statistics are about the same for both and in fact better for the less protected guy.

The necessity felt by modern Indian youth to also earn by climbing trees has mothered the invention of crude but locally manufactured inexpensive tree-climbing machines, as shown in Figure 1.4(c).

1.3.3 INHERITED ETHNIC INFLUENCE

The Americans had one lucky break when they started building skyscrapers which other nations did not have at the time: The unique ability of certain Red Indians – particularly of the Mohawk Tribe, who could walk on any narrow ledge and work at any height without fear or reduced efficiency and hence were called 'Skywalkers'. While one may not attribute this to any special physical gift or genetics, the unique inner sense and skill for heights they had must surely have been passed down the generations of moving to and at height.

FIGURE 1.4

(a) Man climbing coconut palm.
   Source: wikiHow – to Climb a Palm Tree (License: Creative Commons).
   Author: wikiHow Staff. Link: www.wikihow.com/Climb-a-Palm-Tree
(b) American worker cutting tree. [With permission from: Rent A Monkey Tree Service, Sandy, Utah, USA. Website: www.rentamonkey.com]
(c) Coconut Tree climbing machine. [Source: Own work. Author: Manojk. Link: https:// commons.wikimedia.org/wiki/File:Coconut_tree_climbing_DSCN0222.jpg]

Most of the early skyscrapers and bridges around Montreal and New York were built mainly with Mohican support.

Figure 1.5(a,b) displays a Mohawk Indian worker on top of a steel column, with the Chrysler Building in the background, and two others climbing a tower; (c) shows the Mohawk Warrior Flag with the famous Mohawk head, and (d) the 2015 issue of the Native American $1 Coin Program series, one side showing the Mohawk ironworker [Ref. **1.3**].

1.3.4 CULTURAL FACTORS

Although cultural differences are being gradually being 'levelled' out by the Internet, there still remain some major variations in the matter of safety.

For instance, the West with many of its civilisations younger than the East's still retains the spirit of adventure, the 'why not?' and 'make-do' attitudes that their earlier generations had. People are willing, ready, and able to seek out and face new challenges, by striving to do well and competing to win in them, or just to experience the thrill of doing something daring for its own sake.

But in the East, probably subservient to colonial powers and invaders in previous centuries, and/or held back by their more conservative philosophies, people have become used to (and are generally content with) being told what to do and how to do it.

FIGURE 1.5   The Mohawk influence.

(a) Mohawk Skywalker atop a steel column and
(b) Mohawk Indians working on skyscraper.
[Both pictures (a) and (b), with permission from the website of 'White Wolf Pack'. Link: www.whitewolfpack.com/2012/09/the-mohawks-who-built-manhattan-photos.html]
(c) Mohawk Warrior Flag. [Source: Own work. Author: Xasartha; Link: https://commons. wikimedia.org/wiki/File:Flag_of_the_Mohawk_Warriors%27_Society.png]
(d) 2015 issue of the Native American $1 Coin Program, obverse side showing Mohawk woman Sacagawea carrying her infant son, and the reverse side showing Mohawk ironworkers. [Source: US Mint. Authors: Glenna Goodacre and Ronald D. Sanders. Link: www.usmint. gov/coins/coin-medal-programs/native-american-dollar-coins/2015-mohawk-ironworkers]

Their family ties, undergirded by individual responsibilities to extended families sticking together (until recently), may have bred a cautionary shell around themselves, avoiding risk as far as possible.

Even within the same culture, differences will exist, as between country folk and city dwellers. In most countries, urban culture is more street smart, as for instance taking escalators literally in their stride, while their village brethren would hesitate to step on one. Contrarily, the more resilient villager will casually walk on a fallen tree across a stream while the city dweller may need to be led across.

Certainly, thanks to the computer, Internet, and resulting advances in travel and communication, there have been active exchanges of cultural norms all over the world to the extent that soon most cultures may deal with fall management in a uniform fashion.

When we talk about culture here, we refer to 'safety culture' at the workplace rather than about social behaviour of the workers, that is, more on how aware they are of job-related dangers and how responsive to control measures for them.

The foregoing may explain why at the workplace, immigrant workers have more accidents – hence correspondingly more fall injuries and fatalities – than local workers. According to a paper [Ref. **1.4**] the main reasons for this are: (a) They are being given harder and more dangerous tasks; (b) they are not getting worker rights; and (c) they fear losing the job if they complain of difficulties or sickness on the job.

The bottom line in fall safety at the workplace is that the management is responsible for the personnel under their care to protect them from harm when accessing and working at height – their effectiveness depending on the resources available and the enforcement of regulations applicable. Also, the management must hold the immigrants to the same safety standards and rules as locals.

1.4   STATISTICS OF FALL ACCIDENTS AND DEATHS

1.4.1   GENERAL FALL STATISTICS

There are many measures of a country's health and safety in terms of fatalities. More than the actual numbers, their ratios with some widely – if not universally – accepted base value would be significant for comparison and improvement purposes.

The statistics of fall injuries are staggering. The World Health Organization (WHO) estimated in 2021 that 684,000 fatal falls occurred, making it the second leading cause of unintentional (meaning accidental) death, after road traffic fatalities [Ref. **1.5**]. Over 80% of fall-related fatalities occur in low- and middle-income countries, with regions of the Western Pacific and South East Asia accounting for 60% of these deaths.

In all regions of the world, death rates are highest among adults over the age of 60 years. Children learning to walk and soon after are also at great fall risk.

Annually, another 37.3 million falls are severe enough to require medical attention. Globally, falls are responsible for over 38 million DALYs (Disability-Adjusted Life Years) lost each year and result in more years lived with disability than transport injury, drowning, burns, and poisoning combined.

Display of a lot of worldwide statistics will be 'information overload'; hence, just one partial set of data will be presented to provide a flavour of fall events around the world. Actually, the general trends have not changed much for better or worse in most countries, or over the last few decades.

Figure 1.6 displays results on falls for countries with the least (ten), middle (ten), and most (ten) fall ratios for the year 2013, plotted from data for 183 countries. The complete data set, gathered from WHO and other public domain sources, is available in Appendix A.

Global General Deaths, Unintentional Deaths and Fall Deaths in 2013

No.	Abbr.	Country	Pop., 1000s	All Causes	Unin. Death	Fall Deaths	Fall/ UD%	UD/ AC%	AC/ Pop%
1	AFG	Afghanistan	30683	257478	28884	4170	14.4	11.2	0.84
2	AGO	Angola	23448	345159	27915	1511	5.4	8.1	1.47
3	ALB	Albania	2883	21492	671	59	8.8	3.1	0.75
4	ARE	United Arab Emirates	9040	14108	2025	263	13.0	14.4	0.16
5	ARG	Argentina	42538	328060	15000	1591	10.6	4.6	0.77
6	ARM	Armenia	2992	28464	890	76	8.6	3.1	0.95
7	ATG	Antigua and Barbuda	90	555	35	0	0.0	6.3	0.62
8	AUS	Australia	23271	147281	5268	1937	36.8	3.6	0.63
9	AUT	Austria	8487	79388	2917	880	30.2	3.7	0.94
10	AZE	Azerbaijan	9498	63871	2443	200	8.2	3.8	0.67
11	BDI	Burundi	10466	103891	10568	1209	11.4	10.2	0.99
12	BEL	Belgium	11153	109102	4433	1549	34.9	4.1	0.98
13	BEN	Benin	10322	97703	8258	793	9.6	8.5	0.95
14	BFA	Burkina Faso	17085	159668	13950	1669	12.0	8.7	0.93
15	BGD	Bangladesh	157157	869513	54906	6107	11.1	6.3	0.55
16	BGR	Bulgaria	7253	110452	2038	484	23.8	1.8	1.52
17	BHR	Bahrain	1349	3120	212	18	8.6	6.8	0.23
18	BHS	Bahamas	378	2439	124	8	6.8	5.1	0.65
19	BIH	Bosnia and Herzegovina	3824	38766	1420	149	10.5	3.7	1.01
20	BLR	Belarus	9497	132200	8691	2320	26.7	6.6	1.39
21	BLZ	Belize	344	2040	164	10	5.9	8.0	0.59
22	BOL	Bolivia (Plurinational State of)	10400	70493	7434	445	6.0	10.5	0.68
23	BRA	Brazil	204259	1215409	80759	13875	17.2	6.6	0.60
24	BRB	Barbados	283	3140	97	9	8.7	3.1	1.11
25	BRN	Brunei Darussalam	412	1407	84	13	15.6	6.0	0.34
26	BTN	Bhutan	755	4725	359	62	17.3	7.6	0.63
27	BWA	Botswana	2177	16311	890	34	3.8	5.5	0.75
28	CAF	Central African Republic	4711	71458	4596	210	4.6	6.4	1.52
29	CAN	Canada	35231	246557	10171	4726	46.5	4.1	0.70
30	CHE	Switzerland	8119	64620	2526	1777	70.4	3.9	0.80
31	CHL	Chile	17576	100256	5108	966	18.9	5.1	0.57
32	CHN	China	1400000	9723776	554339	103492	18.7	5.7	0.69
33	CIV	Ivory Coast	22470	281838	23525	2545	10.8	8.3	1.25
34	CMR	Cameroon	22211	241063	19105	2263	11.8	7.9	1.09

(Continued)

Global General Deaths, Unintentional Deaths and Fall Deaths in 2013

No.	Abbr.	Country	Pop., 1000s	All Causes	Unin. Death	Fall Deaths	Fall/ UD%	UD/ AC%	AC/ Pop%
35	COD	Democratic Republic of Congo	72553	735019	58278	2463	4.2	7.9	1.01
36	COG	Republic of the Congo	4394	33887	2539	148	5.8	7.5	0.77
37	COL	Colombia	47342	237709	15166	1610	10.6	6.4	0.50
38	COM	Comoros	752	5788	548	54	9.9	9.5	0.77
39	CPV	Cape Verde (Cabo Verde)	507	2851	190	11	6.0	6.7	0.56
40	CRI	Costa Rica	4706	21297	1365	103	7.6	6.4	0.45
41	CUB	Cuba	11363	92298	5215	2199	42.2	5.7	0.81
42	CYP	Cyprus	1142	7926	316	47	14.9	4.0	0.69
43	CZE	Czech Republic (Czechia)	10545	109023	3515	672	19.1	3.2	1.03
44	DEU	Germany	80566	892625	22515	11367	50.5	2.5	1.11
45	DJI	Djibouti	865	7126	591	67	11.4	8.3	0.82
46	DNK	Denmark	5624	51973	1215	541	44.5	2.3	0.92
47	DOM	Dominican Republic	10281	60917	4558	128	2.8	7.5	0.59
48	DZA	Algeria	38186	188768	16167	1472	9.1	8.6	0.49
49	ECU	Ecuador	15661	79098	7619	639	8.4	9.6	0.51
50	EGY	Egypt	87614	547485	23331	2969	12.7	4.3	0.62
51	ERI	Eritrea	4999	31683	3036	280	9.2	9.6	0.63
52	ESP	Spain	46455	390246	9572	2681	28.0	2.5	0.84
53	EST	Estonia	1320	15254	411	113	27.4	2.7	1.16
54	ETH	Ethiopia	94558	709594	65336	8053	12.3	9.2	0.75
55	FIN	Finland	5453	51325	1950	1126	57.7	3.8	0.94
56	FJI	Fiji	881	6081	250	18	7.3	4.1	0.69
57	FRA	France	63845	554829	24429	6606	27.0	4.4	0.87
58	FSM	Micronesia	104	645	26	3	10.6	4.0	0.62
59	GAB	Gabon	1650	14431	916	78	8.5	6.3	0.87
60	GBR	U.K. of Great Britain and N. Ireland	63956	575824	13292	5319	40.0	2.3	0.90
61	GEO	Georgia	4083	48572	1375	178	12.9	2.8	1.19
62	GHA	Ghana	26164	230337	18133	2555	14.1	7.9	0.88
63	GIN	Guinea	11949	117900	9551	806	8.4	8.1	0.99
64	GMB	Gambia	1867	15956	1501	181	12.1	9.4	0.85
65	GNB	Guinea-Bissau	1757	18166	1358	132	9.8	7.5	1.03
66	GNQ	Equatorial Guinea	797	8515	676	52	7.7	7.9	1.07
67	GRC	Greece	11055	118549	2665	521	19.6	2.2	1.07
68	GRD	Grenada	106	852	36	6	15.8	4.2	0.80
69	GTM	Guatemala	15691	84752	8712	16	0.2	10.3	0.54
70	GUY	Guyana	761	6198	414	12	2.8	6.7	0.81
71	HND	Honduras	7849	35110	1989	131	6.6	5.7	0.45
72	HRV	Croatia	4272	50306	1905	1041	54.6	3.8	1.18
73	HTI	Haiti	10431	91282	7401	404	5.5	8.1	0.88

Global General Deaths, Unintentional Deaths and Fall Deaths in 2013

No.	Abbr.	Country	Pop., 1000s	All Causes	Unin. Death	Fall Deaths	Fall/ UD%	UD/ AC%	AC/ Pop%
74	HUN	Hungary	9925	126270	3489	1822	52.2	2.8	1.27
75	IDN	Indonesia	251268	1816804	89290	17519	19.6	4.9	0.72
76	IND	India	1300000	9456822	789320	196474	24.9	8.3	0.73
77	IRL	Ireland	4671	29410	710	200	28.2	2.4	0.63
78	IRN	Iran (Islamic Republic)	77152	361357	32124	1956	6.1	8.9	0.47
79	IRQ	Iraq	34107	177874	10915	562	5.1	6.1	0.52
80	ISL	Iceland	325	1995	71	18	26.1	3.5	0.61
81	ISR	Israel	7818	41448	1233	130	10.6	3.0	0.53
82	ITA	Italy	59771	614037	18828	3781	20.1	3.1	1.03
83	JAM	Jamaica	2773	18946	783	11	1.4	4.1	0.68
84	JOR	Jordan	7215	27775	2559	145	5.7	9.2	0.38
85	JPN	Japan	126985	1265658	44280	7956	18.0	3.5	1.00
86	KAZ	Kazakhstan	17100	151872	10592	1121	10.6	7.0	0.89
87	KEN	Kenya	43693	315909	28846	3026	10.5	9.1	0.72
88	KGZ	Kyrgyzstan	5746	36800	2180	107	4.9	5.9	0.64
89	KHM	Cambodia	15079	94457	7960	678	8.5	8.4	0.63
90	KIR	Kirbati	109	766	27	1	4.8	3.6	0.71
91	KOR	South Korea (Republic of Korea)	24896	265336	14417	2541	17.6	5.4	1.07
92	KWT	Kuwait	3594	8958	971	96	9.9	10.8	0.25
93	LAO	Laos (People Democ. Repub.)	6580	48227	3499	315	9.0	7.3	0.73
94	LBN	Lebanon	5287	32524	1916	571	29.8	5.9	0.62
95	LBR	Liberia	4294	36730	3169	298	9.4	8.6	0.86
96	LBY	Libya	6266	30895	2677	268	10.0	8.7	0.49
97	LCA	Saint Lucia	182	1324	77	2	2.6	5.8	0.73
98	LKA	Sri Lanka	20522	138978	10149	2065	20.3	7.3	0.68
99	LSO	Lesotho	2083	27971	1150	39	3.4	4.1	1.34
100	LTU	Lithuania	2964	42367	1521	418	27.5	3.6	1.43
101	LUX	Luxembourg	545	3634	190	58	30.5	5.2	0.67
102	LVA	Latvia	2012	29464	982	198	20.2	3.3	1.46
103	MAR	Morocco	33453	191384	11070	911	8.2	5.8	0.57
104	MDA	Republic of Moldova	4074	44668	1696	218	12.9	3.8	1.10
105	MDG	Madagascar	22925	158065	13794	1015	7.4	8.7	0.69
106	MDV	Maldives	351	1175	85	17	20.4	7.2	0.33
107	MEX	Mexico	123740	597140	36648	2642	7.2	6.1	0.48
108	MKD	Macedonia	2073	19932	404	49	12.1	2.0	0.96
109	MLI	Mali	16592	181414	13573	1193	8.8	7.5	1.09
110	MLT	Malta	417	3249	80	40	50.7	2.4	0.78
111	MMR	Myanmar	52984	433210	30188	7346	24.3	7.0	0.82
112	MNE	Montenegro	625	6402	167	20	12.2	2.6	1.03
113	MNG	Mongolia	2859	19361	1269	180	14.2	6.6	0.68

(Continued)

Global General Deaths, Unintentional Deaths and Fall Deaths in 2013

No.	Abbr.	Country	Pop., 1000s	All Causes	Unin. Death	Fall Deaths	Fall/ UD%	UD/ AC%	AC/ Pop%
114	MOZ	Mozambique	26467	301579	22212	2076	9.3	7.4	1.14
115	MRT	Mauritania	3873	31287	2420	269	11.1	7.7	0.81
116	MUS	Mauritius	1264	9530	354	21	5.9	3.7	0.75
117	MWI	Malawi	16190	156299	10805	1295	12.0	6.9	0.97
118	MYS	Malaysia	29465	144059	10420	636	6.1	7.2	0.49
119	NAM	Namibia	2347	15578	1079	36	3.3	6.9	0.66
120	NER	Niger	18359	172910	14368	1088	7.6	8.3	0.94
121	NGA	Nigeria	172817	2228868	142290	20690	14.5	6.4	1.29
122	NIC	Nicaragua	5946	29290	2032	258	12.7	6.9	0.49
123	NLD	Netherlands	16809	140751	4343	2185	50.3	3.1	0.84
124	NOR	Norway	5084	41022	1673	497	29.7	4.1	0.81
125	NPL	Nepal	27835	179603	10670	1610	15.1	5.9	0.65
126	NZL	New Zealand	4465	30278	1166	549	47.0	3.9	0.68
127	OMN	Oman	3907	10735	1479	96	6.5	13.8	0.27
128	PAK	Pakistan	181193	1366445	88761	1411	1.6	6.5	0.75
129	PAN	Panama	3806	18935	906	170	18.8	4.8	0.50
130	PER	Peru	30566	162660	12362	1269	10.3	7.6	0.53
131	PHL	Philippines	97572	653755	34053	3288	9.7	5.2	0.67
132	PNG	Papua New Guinea	7309	56450	3702	155	4.2	6.6	0.77
133	POL	Poland	38619	387406	14203	5231	36.8	3.7	1.00
134	PRK	Republic of Korea	49847	230027	11545	1648	14.3	5.0	0.46
135	PRT	Portugal	10460	106649	2767	717	25.9	2.6	1.02
136	PRY	Paraguay	6466	34376	2762	295	10.7	8.0	0.53
137	QAT	Qatar	2101	3173	618	82	13.2	19.5	0.15
138	ROU	Romania	19794	249743	6476	1272	19.6	2.6	1.26
139	RUS	Russian Federation	143367	2021294	126960	23929	18.8	6.3	1.41
140	RWA	Rwanda	11078	72789	9097	1242	13.7	12.5	0.66
141	SAU	Saudi Arabia	30201	101957	13256	1276	9.6	13.0	0.34
142	SDN	Sudan	38515	300484	31071	2612	8.4	10.3	0.78
143	SEN	Senegal	14221	92488	8700	856	9.8	9.4	0.65
144	SGP	Singapore	5405	25259	506	202	39.9	2.0	0.47
145	SLB	Solomon Islands	561	3010	244	11	4.4	8.1	0.54
146	SLE	Sierra Leone	6179	85355	6984	770	11.0	8.2	1.38
147	SLV	El Salvador	6090	40279	2666	461	17.3	6.6	0.66
148	SOM	Somalia	10268	148251	10304	745	7.2	7.0	1.44
149	SRB	Serbia	8938	107931	1759	424	24.1	1.6	1.21
150	SSD	South Sudan	11454	128008	9696	871	9.0	7.6	1.12
151	STP	Sao Tome and Principe	182	1203	102	16	15.9	8.5	0.66
152	SUR	Suriname	534	3832	241	19	8.0	6.3	0.72
153	SVK	Slovakia	5419	51963	2177	1239	56.9	4.2	0.96
154	SVN	Slovenia	2065	19247	935	540	57.7	4.9	0.93
155	SWE	Sweden	9624	90237	2873	1000	34.8	3.2	0.94
156	SWZ	Swaziland	1251	12411	707	18	2.5	5.7	0.99

Global General Deaths, Unintentional Deaths and Fall Deaths in 2013

No.	Abbr.	Country	Pop., 1000s	All Causes	Unin. Death	Fall Deaths	Fall/ UD%	UD/ AC%	AC/ Pop%
157	SYC	Seychelles	95	728	40	7	16.8	5.5	0.76
158	SYR	Syrian Arab Republic	19323	156690	5572	337	6.0	3.6	0.81
159	TCD	Chad	13146	178621	13963	1230	8.8	7.8	1.36
160	TGO	Togo	6929	63277	5250	479	9.1	8.3	0.91
161	THA	Thailand	67451	533217	43848	6597	15.0	8.2	0.79
162	TJK	Tajikistan	8112	45576	3802	380	10.0	8.3	0.56
163	TKM	Turkmenistan	5240	40977	2385	186	7.8	5.8	0.78
164	TLS	East Timor (Timor-Leste)	1129	8054	610	57	9.4	7.6	0.71
165	TON	Tonga	105	621	35	4	10.1	5.6	0.59
166	TTO	Trinidad and Tobago	1348	12399	431	36	8.3	3.5	0.92
167	TUN	Tunisia	11006	72794	4228	400	9.5	5.8	0.66
168	TUR	Turkey	76224	438865	21305	4438	20.8	4.9	0.58
169	TZA	Tajikistan	8112	43433	3712	380	10.2	8.5	0.54
170	UGA	Uganda	36573	296728	26148	2841	10.9	8.8	0.81
171	UKR	Ukraine	45165	691833	28434	5590	19.7	4.1	1.53
172	URY	Uruguay	3408	32736	1717	50	2.9	5.2	0.96
173	USA	United States of America	317136	2591371	98775	30584	31.0	3.8	0.82
174	UZB	Uzbekistan	29033	188636	9795	727	7.4	5.2	0.65
175	VCT	St Vincent and the Grenadines	109	906	43	6	12.8	4.7	0.83
176	VEN	Venezuela (Bolivarian Republic of)	30276	164154	17168	982	5.7	10.5	0.54
177	VNM	Viet Nam	91379	527414	48459	11659	24.1	9.2	0.58
178	VUT	Vanuatu	253	1229	80	4	5.2	6.5	0.49
179	WSM	Samoa	190	990	57	5	8.4	5.8	0.52
180	YEM	Yemen	25533	157608	15407	1367	8.9	9.8	0.62
181	ZAF	South Africa	53417	574470	23899	1169	4.9	4.2	1.08
182	ZMB	Zambia	15246	130914	9714	922	9.5	7.4	0.86
183	ZWE	Zimbabwe	14898	144058	11409	411	3.6	7.9	0.97
	Averages		39094	301816	18900	3389	15.7	6.2	0.80
	MAX		1400000	9723776	789320	196474	70.4	19.5	1.53
	MIN		90	555	26	0	0.0	1.6	0.15

The results should be viewed in proper perspective: The data is invariably what is supplied by the various countries, each of which will have its own ways of observing, measuring, and recording fall data, each step of which may be subject to human, technical, administrative, and even political biases and errors. Additionally, for countries from which data has not been received or is incomplete, analysts often assume rates prevailing in 'similar' countries.

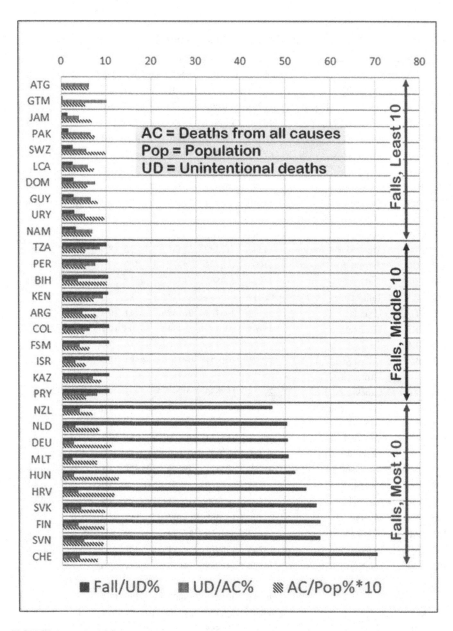

FIGURE 1.6 Fall fatality rates, lowest 10, middle 10, and highest 10.

Source: Raw numeric data from WHO and ILO and other public domain websites. Author: Own work; ratios computed and plotted by Natarajan Krishnamurthy.

However, certain overall conclusions may be extracted from the chart, as follows:

1. Ratios of deaths from all causes to population vary from 0.15% to 1.5%, with an average of 0.8%, quite stable for all countries.

2. Ratios of unintentional, that is accidental, deaths to total deaths vary from 1.6% to 20%, with an average of 6%, fairly stable for all countries.
3. Ratios of fall deaths to accidental deaths vary widely from close to zero to 70%, with an average of 16% for all countries. Obviously, fall consequences are random, unpredictable, and not too controllable.

1.4.2 WORKPLACE FATALITY STATISTICS

Workplaces have special differences from general public activity, namely:

(a) They involve adults who are physically and mentally quite able, earning their livelihood by working;
(b) They often involve hazards of varying nature and risk category;
(c) Hence, all personnel at workplaces will come under formal safeguards, particularly safety Acts and Regulations by the Government and safety Codes of Practice and guidelines by expert committees.

Statistics gathered from workplaces are, therefore, more reliable and revealing than general fall statistics.

A widely accepted measure of workplace safety is the fatality rate, taken as the number of deaths per 100,000 full-time workers and full-time equivalents of part-time workers, that is, (number of workplace fatalities) × 100,000/(number of full-time worker equivalents at the workplace).

The number 100,000 was globally accepted as one which, when multiplied by the ratio of workers dead to total number of full-time workers, would yield a result that can be remembered and communicated easier than the straight ratio of the two quantities involved.

Thus, if a country has 48 workplace deaths from a workforce of 1.5 million in a certain year, its workplace fatality rate is (48/1,500,000) = 0.000032. However, using the standard base of 100,000 workers, we get the rate as (48 × 100,000/1,500,000) = 3.2, so much neater!

In 2018, the country with the lowest workplace fatality rate (for 100,000 workers) was Norway with 0.42. Germany had 0.55, the UK 0.61, Sweden 0.66, Italy 1.04, Spain 1.49, and France 3.07; the European Union (EU) as a whole claimed 1.27 [Ref. **1.6**].

Outside EU, for 2017–2018, workplace fatality rate in Singapore was 1.2, Canada 1.9, Australia 2.8, Japan 2.0, Russian Federation 5.0, the USA 5.3, Myanmar 5.9, Malaysia 6.3, and Mexico 11.0.

Again, it may be noted that the values vary widely, basically depending on the safety culture and resources available to the country. No country of global significance had zero fatalities.

1.4.3 WORKPLACE FALL STATISTICS

To get a comprehensive view of falls, along with falls from heights, we must include injuries from slips, trips, and falls at level also. In most countries, the combined fatalities from height and at level are often the highest or close to it, in construction, followed by manufacturing. As for major injuries, manufacturing takes the lead, construction taking second place.

While fall rates in the general population average 16%, falls at workplaces occur more frequently (averaging 30% to 40%). The rate is higher in hazardous industries; construction for instance has deaths from 30% to 50% of all workplace deaths, with fatality rates for 100,000 workers reaching 30 or more.

In sum, workplace fall injuries and deaths are a major cause of concern and lead to a lot of treatment and compensation expense in billions of dollars across the world.

1.5 PLANNING FOR FALL MANAGEMENT

1.5.1 HEIGHT WHEN FALL CONTROL SHOULD BE PROVIDED

Obviously, we cannot avoid for ourselves or prevent for others falls during the thousand and one things we do in our waking hours.

We will also not be able to prevent children falling while learning to walk, as falling may be part of their learning to walk.

However, old folks falling due to lack of balance can be and are given extra care, with medical and geriatric intervention to monitor and manage their mobility, aimed to prevent their falls as much as possible, and minimise harm if and when they do fall.

Beyond these, we can and must prevent falls for normally functioning adults at home, at work, and while indulging in leisure-time activities.

Setting aside for the moment falls at level, research and experience over decades by many nations have shown that there are very few falls for normal adults at heights less than about 2 m. Hence, where governments and other authorities develop guidelines and regulations for the workplace, they stipulate that fall management shall be done for any work at heights more than 2 m to 6 ft in the USA.

However, if, when, and where there are objects or situations below 2 m which may credibly pose physical or health risk to occupants and workers, safeguards against fall risks shall be implemented.

A case in point is where sharp objects, extremely hot or cold temperature sources, toxic or corrosive chemicals, etc., are stored on the floor, and a person falling at level or below 2 m height may come into contact with these danger sources. A potential fall into water or other liquid (or grain, quicksand, etc.) also comes under this clause.

It will, therefore, be recognised that there is really no specific height below which we can say anti-fall safety measures would not be necessary. Fall management should be provided at any and all heights, depending on the circumstances of each case, specifically to those who need it and/or want it.

1.5.2 CORRELATION BETWEEN HEIGHT OF FALL AND INJURY

It would appear quite logical, and commonsensical, to assume that the higher the fall, the more the injury. Regrettably, it is not always so.

Certainly, higher falls will generate more energy and, hence, when an object lands, the impact force will surely be all the greater.

However, actual experience shows mixed findings. People die just falling at level or from less than a metre climbing a ladder. Many lucky ones have unexpectedly survived falls from great heights. There are also numerous instances where with careful planning, a few have jumped without a parachute from hundreds of metres, wearing just wingsuits or not even those, and walked away without a scratch.

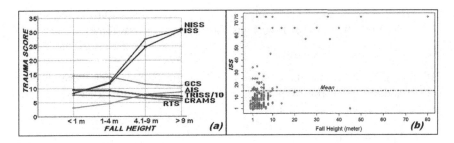

FIGURE 1.7   Correlation between fall height and injuries.

(a) Seven trauma scores versus fall height ranges. [Source: Own work, drawn from Table 3 of Ref. **1.7**.]

(b) Injury Severity Score versus fall height. [Source: Figure 3 of Ref. **1.8**.]

Figure 1.7 displays two sets of data from emergency rooms of hospitals, the plots showing data of standard trauma scores against fall heights.

In Figure 1.7(a), seven trauma scores calculated from the same data from Ref. **1.7** were plotted against height ranges. Three of them show trends of trauma increasing with height, but the other four the opposite! Too, the standard deviations – which are a measure of the reliability of the conclusions – were also quite high, sometimes higher than the mean! Therefore, no firm conclusion could be drawn on the correlation between fall trauma and fall height.

Although the fatality percentages (according to the tabulated values shown) increased with height, injuries for the survivors were horrendous.

In Figure 1.7(b), from Ref. **1.8**, the high values at (or close to) the maximum Injury Severity Score (ISS) value of 75 dot the chart right all the way from about 5 to 80 m.

The fact that most falls occurred within about 10 m may reflect that more persons worked at lower heights than at larger heights.

What is more worthy of note is that fairly large numbers of injury points are clustered near the origin, even at the smallest heights in the range.

Most authors of injury–height data (including me) note that in general falls from greater heights may lead to more injuries, but fall injuries are the result of a complex combination of many other factors, such as surface and elastic properties of the fallen body and the landing surface at the instant of contact, etc. – all of which weaken the predictability of fall consequences in a particular case.

We will further examine the reasons for these anomalies later.

1.5.3   SCOPE OF FALL MANAGEMENT

- Consequences of a fall are not happenstance. There is an established, predicable logic to it. There is a lot of science, and even art, and – done with care – yes, some fun, to falling.
- Most of the time, by understanding what happens when one falls, and then analysing our surroundings, we can avoid falls. With training and proper planning, we can reduce injuries or even evade any harm if and when we fall.

- When we talk about fall height, we will be referring to 'Free Fall', only under the action of gravity and any other related natural forces such as air resistance.

'Fall Management' – sometimes called 'Fall Protection' – is a general term that may cover shielding the person from all fall harm to the maximum extent by whatever means we can devise, as follows

(a) Fall prevention

Preventing a fall is simply ensuring that there is no fall. This would also include the elimination of fall hazard altogether, say by not requiring the person to go to height at all. If persons have to access heights, measures must be taken to prevent them from falling over exposed edges. Only if prevention is not possible, we resort to 'fall arrest'.

(b) Fall arrest

'Arrest' of a fall means stopping the fall before the falling person crashes to the ground or other base, thus protecting him from harm of impact when he reaches the end of the fall and would otherwise contact the landing zone. Rather confusingly, many regulations refer to fall arrest as 'Fall Protection', as alternative to fall prevention; unfortunately, this overlooks the argument that fall prevention is also a protection for the person from the traumatic consequences of falls.

To avoid possible confusion, we will use only the terms 'prevention' and 'arrest' of falls as the two clear options available.

1.5.4 IMPERATIVES FOR FALL MANAGEMENT

- Once a fall starts, it is too late to help the victim. Any safeguards to prevent or arrest a fall must be planned and implemented well in advance of the task.
- There should be one person (or department) responsible for workplace safety (including fall management) and legally accountable for the occurrences and consequences of accidents. This entity may be known as the 'Occupier', namely whoever is in overall charge of the project – which may be the owner, employer, project manager, safety manager, or CEO, as designated by the management formally prior to the commencement of the project. (The term 'employer' may be used as equivalent in this book.)
- Overdependence on and overconfidence in individual protective measures after a fall are both unnecessary and often ineffective. Many do not use them right, and most of the individual safeguards demand too many corequisites.
- Formal fall management must be total for the entire space and for all the time under control, with no exceptions of even the smallest fraction. (This is called '100% tie-off' and will be detailed later.)

1.5.5 HIERARCHY OF FALL MANAGEMENT

- Fall prevention is always better than fall arrest. (Grandma always said, 'Prevention is better than cure', didn't she?) A very good reference on the subject is HSE Report 302, [Ref. **1.9**].

- Collective fall control (for all users of a facility) is better than individual fall control. One may argue that if we take care of all the individuals in a group, we would have taken care of the group. But the effort, time, and hence expense, as well as predictability of outcomes will depend almost entirely on the various individuals, which in turn will imply continuous monitoring and corrective measures for each – not suitable for safety.
- From the two preceding requirements, the hierarchy – decreasing order of effectiveness – of fall management may be defined as:
  - (i) Collective Fall Prevention;
  - (ii) Individual Fall Prevention;
  - (iii) Collective Fall Arrest;
  - (iv) Individual Fall Arrest.

Note that (ii) individual fall prevention and (iii) collective fall arrest are both equal in rank, being combinations of one first priority and another second priority criterion, and so should be treated as of equal rank. However, as prevention is universally recognised as better than cure after the onset of a mishap, individual fall prevention is given a slight edge over collective fall arrest in practice.

These four possibilities will be discussed in succeeding chapters.

REFERENCES FOR CHAPTER 1

1.1 Patowary, K., "Kids risking their lives to go to school", *Amusing Planet*, 24 March 2013. Retrieved on 25 June 2023 from: www.amusingplanet.com/2013/03/kids-risking-their-lives-to-reach-school.html

1.2 Yglesias, M., "Different places have different safety rules and That's OK", *Blog*, April 2013. Retrieved on 8 August 2017 from: www.slate.com/blogs/moneybox/2013/04/24/international_factory_safety.html

1.3 McPike, S., "Daring Mohawk ironworkers featured in the 2015 native American $1 coin design", *Inside the Mint*, USA, 14 January 2015.

1.4 Underwood, E., "Unhealthy work: Why migrants are especially vulnerable to injury and death on the job", *Knowable Magazine*, July 2018. Retrieved on 23 June 2023 from: https://knowablemagazine.org/article/society/2018/unhealthy-work-why-migrants-are-especially-vulnerable-injury-and-death-job#:~:text=The%20power%20imbalance%20between%20employers,the%20University%20of%20California%2C%20Davis.

1.5 ———. *Falls*, World Health Organization, April 2021. Retrieved on 7 June 2023 from: www.who.int/news-room/fact-sheets/detail/falls

1.6 ———. "Comparison with other countries", *HSE*. Retrieved on 10 June 2023 from: www.hse.gov.uk/statistics/european/index.htm

1.7 Turgut, K., and Mehmet, E.S., et al., "Falls from height: A retrospective analysis", *World Journal of Emergency Medicine*, 9(1): pp. 46–50, 2018. Retrieved on 23 June 2023 from: www.ncbi.nlm.nih.gov/pmc/articles/PMC5717375/

1.8 Anantharaman, V., Zuhary, T.M., Ying, H., and Krishnamurthy, N., "Characteristics of injuries resulting from falls from height in the construction industry", *Singapore Medical Journal*, 64(4): pp. 237–243, April 2023. Retrieved on 23 June 2023 from: https://journals.lww.com/smj/fulltext/2023/04000/characteristics_of_injuries_resulting_from_falls.4.aspx

1.9 Cameron, I., et al., *A Technical Guide to the Selection and Use of Fall Prevention and Arrest Equipment*. HSE Books (Research Report 302), 311 p, 2005. ISBN: 0-7176-2948-1.

2 Pursuit of height

2.1 TIMELINE FOR TALL CONSTRUCTIONS

It will take volumes to write even a brief history of work at height, because living and working at height have been with us ever since man appeared on this earth. Even a very brief history of work at height must start with the pyramids. However, focussing on modern structures, Figure 2.1 gives an overview of the tallest buildings in the world in 2020. Even taller structures are being built or in planning stages. The race for height is generally tempered by return on investment and feasibility considerations.

No doubt, the height of construction work has been increasing by leaps and bounds. But, despite improvements and innovations in construction technology and safety management, fall accidents continue to occur even in technologically advanced countries.

It is not the implication here that the taller the structure, the more unsafe it is for anyone to fall. The construction of taller buildings is surely much more complex and time-consuming, but the safety aspects – and consequences of falls to humans – are not that much different between a three-storey building and a 30-storey building.

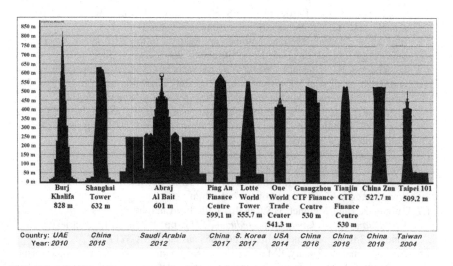

FIGURE 2.1   World's top ten tallest buildings, 2010.

[Source: Data from Emporis.com. Author: Ali Zifan. (Modified for readability). Link: https://commons.wikimedia.org/w/index.php?curid=41356641]

     DOI: 10.1201/9781032648132-2

What is more important for us here is the workplace safety aspect of high-rise construction and other activities, covering the dangers in various tasks involved in life and work at height, their consequences, and ways and means of managing the dangers.

What follows is a brief timeline of life and work at heights from millennia past. We will look at a few highlights of some of these and other tall structures in chronological order and seek lessons from their history.

The lists to follow, all in chronological order and in generally increasing heights, are merely a sampler of the many height-related architectural and engineering achievements since the beginning of recorded time. No claim is made for completeness or sampling logic in these lists. While most of the information is available in public domain, particular mention must be made of detailed tabulated summaries in Wikipedia [Ref. **2.1**].

All years are in the Common Era (CE, formerly AD), with those earlier being marked Before Common Era (BCE, formerly BC). Where exact year is not known, estimated year is prefixed with 'c.' (for 'circa').

Figure 2.2 is a montage of some of the more recognisable landmark structures from the above list.

2.1.1 TALL RELIGIOUS/SPIRITUAL STRUCTURES

Religion-oriented structures took the lead in building tall.

c. 2650 BCE	Pyramid of Djoser	Egypt	63 m
c. 2570 BCE	Great Pyramid of Giza	Egypt	147 m
537–1880	Many Cathedrals	Europe, UK	55–160 m
1010	Brihadhiswarar Temple	Tanjore, India	66 m
1311	Lincoln Cathedral	Lincoln, England	160 m
1993	Hassan-II mosque	Casablanca, Morocco	210 m

2.1.2 TALL BUILDINGS

Buildings are the most iconic and visible tall structures.

1908	Singer Building	New York, USA	186 m
1913	Woolworth Building	New York, USA	241 m
1930	Chrysler Building	New York, USA	320 m
1931	Empire State Building	New York, USA	443 m
1969	John Hancock Center	Chicago, USA	459 m
1978	1-World Trade Center*	New York, USA	526 m
1998	Petronas Twin Towers	KL, Malaysia	452 m
2004	Taipei-101	Taiwan	508 m
2010	Burj Khalifa	Dubai, UAE	830 m

[* Destroyed in terrorist attack in 2001, rebuilt in 2014 to 541 m height.]

FIGURE 2.2 Sampler of world's tallest structures:

Row 1 – Pyramid of Giza, Qutb Minar, Washington Monument, Tokyo Skytree;
Row 2 – Statue of Unity, Burj Khalifa, Tower Bridge, Millau Viaduct;
Row 3 – CN Tower, Golden Gate Bridge, Hoover Dam, Eifel Tower;
Row 4 – Brooklyn Bridge, Sidu River Bridge;
Row 5 – Petronas Towers, Italia Bridge, Brihadhiswarar Temple, Empire State Building.
[Source: Various travel brochures and websites in public domain. See lists for locations and dates.]

2.1.3 TALL BRIDGES

Bridge heights are from deck to foundation level.

1883	Brooklyn Bridge	New York, USA	84 m
1894	Tower Bridge	London, UK	65 m
1929	Royal Gorge Bridge	Canon City, USA	291 m
1937	Golden Gate Bridge	San Francisco, USA	227 m
1998	Akashi Kaikyo Bridge	Japan	298 m
2004	Millau Viaduct	France	336 m
2009	Sidu River Bridge	China	496 m

2.1.4 TALL DAMS

Dams have been in use from prehistoric times because of the need for water in scarce times and for irrigation purposes. A few examples noteworthy for their height follow.

1936	Hoover Dam	USA	222 m
1963	Bhakra Dam	India	226 m
1964	Grand Dixence Dam	Switzerland	285 m
1980	Nurek Dam	Tajikistan	304 m
2013	Jinping-1 Dam	China	305 m

2.1.5 TALL MONUMENTS AND TOWERS

Rulers and victorious powers have erected tall monuments and towers to memorialise their rule or mark their victory or simply to remember a great personage. A selection follows:

c.1200	Qutb Minar	Delhi, India	73 m
1884	Washington Monument	DC, USA	169 m
1889	Eiffel Tower	Paris, France	324 m
1967	Ostankino Tower	Moscow, Russia	540 m
1976	CN Tower	Toronto, Canada	553 m
2011	Tokyo Skytree	Japan	634 m

2.1.6 TALL STATUES

Statue heights are total heights including pedestal.

c. 280 BCE	Colossus of Rhodes*	Greece	30 m
803	Leshan Giant Buddha	Sichuan, China	71 m
1886	Statue of Liberty	New York, USA	93 m
1981	Motherland Monument	Kyiv, Ukraine	102 m
2008	Spring Temple Buddha	Lushan, China	208 m
2018	Statue of Unity	Gujarat, India	240 m

[*Destroyed by earthquake in 226 BCE.]

2.2 HISTORICAL EXAMPLES OF TALL CONSTRUCTION

Regardless of when something had been built or fabricated, we need to examine what (if any) considerations of workplace safety existed at those times. The historical evolution of safe methods of construction and fabrication would serve to help manage the present and improve the future of workplace safety.

Although there are few documents on how the older tall structures were erected, analysis of available data and historical sites have demonstrated the prevalence of a fairly sophisticated use of the materials and application of the technology of the times.

Paintings of these edifices by later artists have also depicted at least the fabrication and erection techniques of the era of the artists – augmented by their own imagination based on whatever scriptural or other descriptions they might have accessed – if not of the historical period of the structure itself.

Fortunately, from the early twentieth century, ample photographic and other documented records give us a very good idea of the construction practices and fall management measures.

A few outstanding examples of construction practices, relevant to work at height, will be given here.

2.2.1 ARTISTS' VISUALISATIONS OF ANCIENT CONSTRUCTION TECHNIQUES

In each of the four examples of historical evidence to follow, the paintings of constructions clearly show that the builders achieved their aims admirably, despite the lack of the site safeguards such as guardrails and personal protective equipment such as helmets that we routinely use today.

Obviously, there would have been accidents leading to injuries and deaths during such arduous tasks involving tonnes of stones and huge quantities of other materials and thousands of workers with only primitive tools to supplement their individual and collective physical strength. As there remain no records on the matter, it may be inferred that there was no incentive or pressure to document any work-related accidents, let alone measures to control the risks that would lead to them.

This apparent lack of concern for the human element may be attributed to the difficulty of documentation in those times and the ignorance of the hazards of working with heavy objects or even of the need to consider workplace accidents and workers' injuries as worthy of record.

We must also note that man's sensibilities and social mores have not always been what we accept as fair and normal today. In ancient society, workers, servants, and slaves were simply tools for their employers and masters to get done difficult, dirty, or tedious tasks which they could not or did not wish to do themselves. Moreover, the group or individuals in power 'owned' everything and everybody so that nobody could cross them.

On the plus side, most religious beliefs extolled 'humane' (even if not equal) treatment of all. There is also evidence to show that the workers were not all slaves and were reasonably well treated, with even fractures set and treated.

Figure 2.3 displays four cases of artists' impressions of ancient construction methods.

FIGURE 2.3 Artists' impressions of ancient constructions.

(a) Pyramids of Egypt. [Source: With permission from Ask Aladdin Travel Experts, Link: www.ask-aladdin.com/Pyramids-of-Egypt/]

(b) Tower of Babel, by unknown artist. [Source: Web Gallery of Art. Author: Unknown German artist. Link: https://commons.wikimedia.org/wiki/File:Tower-of-babel-unknown.jpg]

(c) Tower of Babel, by unnamed artist. [Source: Book 'Bedford Hours'. Author: Meister der Münchner Legenda Aurea. Link: https://commons.wikimedia.org/wiki/File:Building_of_the_Tower_of_Babel_-_Bedford_Hours_%281414-1423%29,_f.17v_-_BL_Add_MS_18850.jpg

(d) Tower of Babel, by Breughel. [Source: Sotheby's of London. Author: Pieter Breughel the Elder. Link: https://commons.wikimedia.org/wiki/File:Pieter_Bruegel_the_Elder_-_The_Tower_of_Babel_(Vienna)_-_Google_Art_Project_-_edited.jpg]

(e) Detail of scaffold from (d). [Enhancement of (d), from Wikimedia commons.]

(f) Detail of treadwheel crane from (d). [Enhancement of (d), from Wikimedia commons. Link: https://en.wikipedia.org/wiki/Treadwheel_crane#/media/File:Tretkran_(Bruegel).jpg]

(g) Roman viaduct. [Source and Author: TheCollector.com, with permission from the Author. Link: www.thecollector.com/how-did-romans-build-aqueducts/]

(h) Cupola of Tanjore Big Temple. [Source: Quora; Author: Unknown: Link: www.quora.com/How-did-Cholas-lift-80-tons-of-Granite-to-a-216-foot-tower-Gopuram-at-the-Tanjore-Brihadeeswarar-Temple]

While the devices shown in the figure might not have been the actual devices presumed in the myth, the painting would certainly reflect the construction practices of the times in which the artist lived.

None of the pictures show a single guardrail or helmet – neither the management nor the workers felt any need for them!

(a) Pyramids of Egypt (c. 2500 BCE)

We cannot visualise how the massive pyramids, with each stone weighing around 2 tons, were moved to the locations and heights of the pyramids of Egypt, some 4,500 years ago, when even the wheel had not been invented, in the absence of the materials, equipment, and high-tech skills we take for granted today!

Without the wheel, the Egyptians wetted the soil to lubricate the movement of huge stone blocks on sleds pulled my men and animals, and moved on ramps by direct pull up the slope, as shown in Figure 2.3(a), as well as pull down by ropes going round stakes ahead to serve the same function as pulleys to move the blocks,

In addition to ramps, they were also quite smart in the use of ladders to reach heights and working there on scaffolds,

(b) Tower of Babel (c. 2200 BCE)

The Tower of Babel, myth or real, said to be built by tribes to reach the heavens, has fascinated artists for decades and so we have many examples of its construction as visualised by various artists.

Figure 2.3(b) shows the painting of the Tower of Babel by an unknown artist in the 1590s, displaying many steep ladders to climb and work from. Featured near the right edge of the picture is a very useful device used widely in those eras, known as the 'Treadwheel Crane' to move loads to required heights. It consists of an open wheel of timber, some 3 m in diameter, within which one or more workers push down on the wheel slats around the circumference, which winds or unwinds a cable around a drum to raise and lower a load from the end of a jib.

Figure 2.3(c) is a picture of the Tower by one of the artists illustrating sacred books, during 1414–1423. It shows a two-stage lifting operation by hand-operated winches. What is intriguing – and preoccupying the artist's mind – is the man falling down from the topmost level on the left of the building, with one lone worker at the bottom craning his neck to see him. This is proof that falls were not uncommon; were they taken by the others in their stride?

Figures 2.3(d), (e), and (f) make a series. The leftmost is a 1563 painting of the Tower by Pieter Bruegel the Elder [Ref. **2.2**], depicting the artist's perception of construction methods of Biblical times. It is a huge canvas and the detail is astounding. Two areas of this painting are shown enlarged in the subsequent two pictures, (e) of a scaffold accessed by a ladder, and (e) of a treadwheel used to lift a load.

(c) Roman viaduct (c. 300 BCE)

Figure 2.3(g) is a clear rendering of the techniques used by Romans for erecting their high viaducts. The scaffold sports cross-bracing, a need possibly discovered after sway collapses. The treadwheel crane drawing is so clear that the three (or more) men inside can be seen. The estimated time span is from about 300 BCE to 200 CE.

(d) Brihadhiswarar temple (1009 CE)

Figure 2.3(h) illustrates a unique event relating to work at height from India. It is about the lifting, by about 66 m (216 ft), of the 80-ton granite top cupola of the Brihadhiswarar (Siva) Temple at Tanjore in South India some centuries ago. It is a UNESCO World Heritage Site, one of the most sacred pilgrimage centres for Hindus, as well as a popular tourist attraction. It has been concluded that they got the cupola dragged by elephants up a 6-km-long ramp.

2.2.2 RECORDS FROM ACTUAL CONSTRUCTIONS

Fortunately, from the early part of the twentieth century, credible records have been maintained on constructions of novel structures including skyscrapers and high bridges. Nowadays, no workplace almost everywhere in the world can function without documentation of compliance with safety regulations. Two of the early pioneering scenarios are presented in Figure 2.4.

The installation in 1857 of the first elevator (termed 'lift' in Asia) in the Haughwout Building in New York City heralded the birth of the 'skyscraper', destined to become the residence and office of choice in cities, eliminating the need to climb scores of steps for each floor.

(a) Empire State Building, New York, USA (1930–31)

The 102-storey Empire State Building, completed in 1931, although not the very first 'skyscraper' – half a dozen steel-framed buildings over 21 floors had already claimed that title, with the 319 m-tall Chrysler Building having just been completed in 1930 – is the one best known and most popular tourist attraction around the world [Ref. **2.3**].

It is also one of the most documented constructions in history, with innumerable photographs (available on the Internet, Ref. **2.4**) of workers moving and fixing massive steel columns and girders, unmindful of how precariously high they were working.

It is fascinating, even humbling, to recollect that it was completed within budget and before scheduled date. Even more importantly, with no safeguards or personal protective equipment of any kind, only five of them died during the construction, just one of them actually by falling from height! The just then completed Chrysler Building reported no workplace deaths! (Reasons for their success have been described in the previous chapter.)

Figure 2.4 (a) is a recent photo of the building. Figure 2.4(b) depicts workers at lunch; although obviously posed, it is simply a dramatisation of the reality of how easily the workers faced and managed the enormous risk of working at height!

(b) Golden Gate Bridge, San Francisco, USA (1937)

The Golden Gate Bridge is worthy of note here, not because of its height record, but for the many innovative solutions that were devised right from the underwater foundation to the construction of the massive towers, and the spinning of the 1 m diameter suspension cables, until the completion of the road surface with its expansion joints to accommodate the nearly 1 m thermal expansion.

FIGURE 2.4 Pioneering tall constructions.

(a) Empire State Building. [Source: Own work. Author: Axel Tschentscher. Link: https://commons.wikimedia.org/wiki/File:Empire_State_Building_From_Rooftop_2019-10-05_19-11_(cropped).jpg]

(b) Steel workers having lunch. [Source: https://petterssonorg.files.wordpress.com/2013/01/rockefeller-center-1932.jpg. Author: Photographer Charles Clyde Ebbets. Link: https://commons.wikimedia.org/wiki/File:Lunch_atop_a_Skyscraper.jpg]

(c) Golden Gate Bridge worker wearing leather hard hat. [Source: Screenshot from original movie footage. Source: Golden Gate Bridge, Highway & Transportation District. With permission from Marketing & Communications Director.]

(d) Chief Engineer Strauss of Golden Gate Bridge. [Source: Own work. Author Attribution: © Steven Pavlov/http://commons.wikimedia.org/wiki/User:Senapa/CC BY-SA 3.0. Link: https://commons.wikimedia.org/wiki/File:Joseph_Strauss_Memorial.jpg]

(e) Safety net under Golden Gate Bridge. [Source: Screenshot from original movie footage. Source: Golden Gate Bridge, Highway & Transportation District. With permission from Marketing & Communications Director.]

What is relevant to this book is the fact that many of the workplace safety hazards that we use today were developed and used during the construction of this bridge, Even the common hard hat of today had its beginnings in the head-protective leather hat invented by Bullard and mandated for all personnel by Chief Engineer for the Golden Gate Bridge Joseph Strauss. Figure 2.4(c) shows a worker proudly wearing the hat, ready for a demonstration of its efficacy by having it hit by an iron rod – what a jump in workplace safety from the workers of the skyscrapers in the adjacent picture!

Chief Engineer Strauss, pictured in Figure 2.4(d), and his colleagues introduced many other workplace safety measures [Ref. **2.5**], most of them aimed at fall management, such as the waist belt work restraint.

Most important among these innovations was the safety net used for the first time in construction to catch personnel falling off their high perches. Pictured in Figure 2.4(e), the net was suspended under the bridge floor during the construction of the roadway structure with stiffening truss along the entire span from pylon to pylon. The $130,000 net, made of manila rope square mesh extended 10 ft (3.0 m) outside the trusses on both sides caught falling workers and gave them confidence to work more quickly without fear of injury even if they fell.

2.3   HISTORY OF FALL REGULATIONS

2.3.1   SAFETY AWARENESS

With the industrial revolution from late 1700s to early 1800s came many improvements to the quality of human life. Buildings began to grow vertically, as urbanisation took over to cater to the industrialisation.

The lack of accident information here was no more the ignorance of the consequences of advancing technology, but has been attributed to ambition of the employers, combined with inherent negligence and exploitation of the system [Ref. **2.6**].

Public gradually became aware of the impact of work at height dangers and the need for regulation and guidelines. What follows are significant landmarks in regulations concerning workplace safety in general and fall safety in particular.

2.3.2   TIMELINE FOR WORKPLACE SAFETY REGULATIONS

Considerable information on the evolution of workplace safety regulations is available [Ref. **2.7**]. An overview of significant regulatory events and developments is as follows.

1802:

UK enacts the Health and Morals of Apprentices Act of 1802, precursor of the Factories Act.

1833:

Health and Safety Executive (HSE) of the UK establishes the Factory Act to prevent injury and overworking in child textile workers [Ref. **2.8**]. It has been steadily expanding over the decades to cover all health and safety aspects of the workforce in all industries.

1837:

The UK court admits and enforces the concept of 'Duty of Care" for employers at workplaces.

1857:

Scientific investigations initiated to detect workplace hazards and unionisation by the workers themselves for self-protection started and combined to create the workplace safety movement.

1948:

Independent India introduces the Factories Act, based on the earlier regulations by the British rulers, covering various aspects of the safety and health of working men, women, and children. Many additions and modifications are made in subsequent decades.

1964:

Canada, spurred by Ontario Government, enacts the Industrial Safety Act in 1964.

1970:

The Occupational Safety and Health Administration (OSHA) is established in the USA to ensure safe and healthful working conditions for working men and women.

Because of a large number of personnel involved and the huge research and development effort invested into it, the OSHA today stands among leading workplace safety and health organisations around the world.

1972:

The OSHA introduces Construction Safety Standards in which requirements for safety belts, lifelines, and lanyards, as well as safety nets were described.

1974:

Health and Safety Executive (HSE) of the UK introduces the Health and Safety at Work Act, 1974.

1880:

The UK enacts Employers' Liability Act.

1974:

The UK passes the comprehensive Health and Safety Act of 1974.

1990:

The OSHA proposes new rules focusing on trips, slips, and falls, detailing requirements for personal protective equipment (PPE) including fall arrest systems, specifically harnesses and deceleration devices (i.e. shock absorbers). Public comment was sought, with intensive interaction with concerned industries and stakeholders.

1992:

The Commission of European Communities conducts a survey of Europeans on Health and Safety at work, which reveals the high importance of falls (71%) and falling objects (53%) [Ref. **2.9**].

1993 to Present:

The European Union (EU) is formally established by the Maastricht Treaty. By 2013, it had 28 members, but in 2016, the UK expressed its intention to leave the EU.

The EU comes up with its own standards on PPE for work at height [Ref. **2.10**].

1994:

The OSHA issues Final Rule on '*Safety Standards for Fall Protection in the Construction Industry*', completely revising regulations for fall protection systems and procedures. These systems and procedures are intended to prevent employees from falling off, onto or from working levels, and to protect employees from falling objects.

1996:

HSE issues '*The Construction (Health, Safety and Welfare) Regulations 1996, No. 1592 Regulation 6*', detailing requirements for various elements of the construction industry, including management of falls of persons and objects.

1996:

The OSHA issues Final Rule on '*Safety Standards for Scaffolds Used in the Construction Industry*', updating standards and setting performance-oriented criteria to protect employees from falls, falling objects, and other hazards.

2001:

In recognition of the highest fall accident and fatality rates in the steelwork, the OSHA issues Final Rule on '*Safety Standards for Steel Erection*'.

2003:

The OSHA reopens the 1990 proposals with revisions to reflect the fast-changing technology in fall protection and again sought public comment.

From the comments received, many issues needed further deliberation and the OSHA institutes a complete overhaul of fall protection standards.

2005:

HSE releases '*Work at Height Regulations – 2005*', a very comprehensive set of recommendations and requirements for working safely at height. Certain modifications are made in 2007 through a supplement titled '*Work at Height (Amendment) Regulations – 2007*'.

2009–2014:

The American National Standards Institute and the American Society of Safety Engineers issue Z359 Fall Protection Code Package v3.0 [Ref. **2.11**], a very comprehensive set of regulations and recommendations for all aspects of fall management.

2015:

The OSHA releases the definitive publication '*Fall Protection in Construction*' to summarise various key aspects of the topic [Ref. **2.12**].

2021:

The Government of India constitutes an expert committee of industry and subject experts to update the Factories Act of 1948 to meet the current requirements meeting technological progress and to increase productivity in the country. (I was privileged to be invited and to participate in the committee's work.)

2.4   HISTORY OF FALL SAFETY

As awareness of workplace dangers improved, measures were innovated, tried, and, after confirming they worked as expected, enforced into codes of practice. The following is an overview of significant events in fall safety.

1919:

Bullard invents the 'Hard hat' [Ref. **2.13**]. Bullard company continuously improves it from 1938 to 1982 to current status.

1933–1937:

The safety measures introduced or enforced by Golden Gate Bridge project Joseph Strauss and his colleagues (already mentioned and illustrated) were quite novel at the time but almost universally adopted today [Ref. **2.5**].

1934:

Clarence W. Rose – who early in his career was a window washer – pioneers fall protection by starting the Rose Mfg. Co., producing safety belts and lanyards (made of rope from natural fibres such as manila hemp) and simple body belts with no shock-absorbing properties, for window washers [Ref. **2.14**].

1959:

Rose patents a cable connector with the first shock absorber to reduce the shock to the wearer when he reached the end of the cable in a fall [Ref. **2.14**].

1968:

Out of necessity, mountaineering, which grew as a sport from the early 1800s, develops a number of safeguards for fall arrest commercially to replace the 'Swami-belt' – the (East) Indian word '*swami*' referring to a Hindu ascetic usually wearing a loincloth – made by wrapping a rope (or, for mountaineering, a woven web) a number of times round the body and knotting between the legs and around the waist.

It is similar to the 'pelvic harness' of today and has elements similar to the safety harness as we know it now.

1970s–1980s:

The anchored rope from the waist belt certainly saved fallen persons from hitting the ground, but more often than not, it damaged their intestines and often injured the spine from the jerk at the end of the rope or cable, because the entire impact force had to be resisted by the backbone, which is quite weak in bending.

The search for a better alternative ends with the development of the full-body safety harness with shock absorber, in the early 1970s. It came into recommended and preferred use in the 1980s and became the mandatory requirement for fall arrest by late 1980s.

1978:

Rose patents the double (or 'twin-tail') lanyard for climbing ladders, each clamp of the pair to be attached alternately to higher rungs as one climbs.

This established and implemented the principle '100% tie-off', which required that a worker's body belt could be secured by two lanyards to two separate anchors, so that while transferring from one location to another, the worker would be assured fall protection by at least one connection, thus for 100% of the length and 100% of the time, affirming its name.

Simple in concept and very inexpensive, it is applicable to all situations where the worker has to negotiate gaps with no other fall prevention measures. But its efficacy is solely dependent on the worker's understanding and acquiescence.

1998:

The waist belt is banned for fall arrest in the USA. Full-body safety harness is mandated in OSHA regulations [Ref. **2.15**].

2000–Present:

The Internet of Things, PPVC, three-dimensional (3-D) printing, Virtual Reality, and other recent technologies, are helping to improve identification, assessment, and control of fall risks in a big way – these will be covered in a later chapter.

REFERENCES FOR CHAPTER 2

2.1 ———. "List of tallest buildings and structures", *Wikipedia*. Retrieved on 12 June 2023 from: https://en.wikipedia.org/wiki/List_of_tallest_buildings_and_structures

2.2 ———. "The tower of Babel (Bruegel)", *Wikipedia*. Retrieved on 19 July 2016 from: https://en.wikipedia.org/wiki/The_Tower_of_Babel_(Bruegel)

2.3 ———. "The Empire State building", *Wikipedia*. Retrieved on 19 July 2016 from https://en.wikipedia.org/wiki/Empire_State_Building

2.4 ———. *Photographs of the Empire State Building under Construction*. New York Public Library Digital Collections. Retrieved on 19 July 1015 from: http://digitalcollections.nypl.org/collections/photographs-of-the-empire-state-building-under-construction#/?tab=about

2.5 Kerievsky, J., "Golden gate safety", *Industrial Logic*, 29 January 2015. Retrieved on 27 June 2023 from: www.industriallogic.com/blog/goldRef. gate-safety/

2.6 Abrams, H.K., "A short history of occupational health", *Advances in Modern Environmental Toxicology,* 22: pp. 33–71, 1994.

2.7 Cameron, D., "History of workplace safety and health", *Health & Safety*, Stay-Safe, London. Retrieved on 16 June 2023 from: https://staysafeapp.com/blog/history-workplace-health-and-safety/

2.8 ———. *The 1833 Factory Act*. UK Parliament. Retrieved on 19 July 2023 from: www.parliament.uk/about/living-heritage/transformingsociety/livinglearning/19thcentury/overview/factoryact/

2.9 ———. *Europeans and Health and Safety at Work – A Survey*. Commission of the European Communities, Luxembourg, 1992.

2.10 ———. *European Commission Single Market Standards: Personal Protective Equipment*. Directive 89/686/EEC, 2016.

2.11 ———. "ANSI/ASSE Z359 – Fall protection code package", *American National Standards Institute*. Retrieved on 27 June 2023 from: https://blog.ansi.org/ansi-asse-z359-fall-protection-code-package/#gref

2.12 ———. *Fall Protection in Construction*. OSHA 3146–05R, USA, 2015. Retrieved on 24 July 2016 from: www.osha.gov/Publications/OSHA3146.pdf

2.13 McLoud, D., "The history of the hard hat . . .", *Equipment World* by Randall Reilly, 24 March 2019. Retrieved on 28 June 2023 from: www.equipmentworld.com/workforce/safety/article/14970852/the-history-of-the-hard-hat-in-the-construction-industry

2.14 Barrera, M., "PPE-volution", *Occupational Safety & Health*, 1 January 2007. Retrieved on 27 June 2023 from: https://ohsonline.com/Articles/2007/01/01/PPEvolution.aspx

2.15 ———. *Fall Protection in Construction, OSHA 3146*. US Department of Labor, 1998 (Revised).

3 Collective fall prevention

3.1 FALL PREVENTION ALTERNATIVES

As collective management is better than individual management, and fall prevention is better than fall arrest, collective fall prevention is the highest priority in fall management.

Collective fall prevention may be achieved by four options:

(a) Avoid activity involving work at height if and as possible;
(b) Simply ban or close access to the height (or depth) risk;
(c) Implement alternative methods circumventing human work at height;
(d) Provide edge protection against falling over any edge at height.

Not going to heights does not stop one from falling at level. To avoid such zero-height falling by slips, trips, or otherwise, we would have to have a walker or other mechanical (or electrical) aid to support the person all the time, which would be feasible only for those who are (or when) physically challenged or sick and not for the general population of the workforce!

3.2 FALL ELIMINATION

3.2.1 AVOIDANCE OF ROUTINE HEIGHT ACTIVITY

Elimination is like saying, 'If you cannot swim, don't go near the water!"

Eliminate activities that require accessing or work at height. Note that this is not strictly fall prevention but fall avoidance, but the outcome is the same.

If you are an employer, think when, where, and how you can avoid your workers or staff going to or working at heights and still achieve your overall objectives for the project. Then, there is no risk of their falling and getting injured!

How can it be done?

Well, it is already being done in a few interesting ways.

(a) Long-handled telescoping tubes and rods with fittings for (i) water jet cleaning, as shown in Figure 3.1(a), (ii) sloping roof-gutter clearing, (iii) window washing, (iv) wiping or painting wall surfaces, etc., are already available commercially to facilitate household chores without undergoing risks of ladder climbing and working at height. (This is limited to two storied buildings.)
(b) Tall lamp posts may be hinged within 1.5 m above the ground so that when bulbs need to be changed, the top portion may be swung down to ground level for it, and then swung back up and locked in position for use. Figure 3.1(b) shows the technique used, with the swung portion temporarily supported on

DOI: 10.1201/9781032648132-3

a trestle T (or by a helper). The inset of hinge H shows the enlarged view of how the electrical connection is safely maintained during the operation.

(c) Standard guardrails may be attached to work platforms and the assembly raised into position by hoist, avoiding workers being exposed to fall hazard while fixing guardrails to platform planks after they are put up, as suggested in Figure 3.1(c).

3.2.2 ELIMINATION OF CONSTRUCTION ACTIVITY AT HEIGHT

Normally, to erect bridges or multiple floors of a building, we use a scaffold for work at different heights from the ground or other base. In the case of reinforced concrete building, we use in addition temporary formwork to cast the concrete and allow it to cure. All of these pose serious hazards of falling from height.

For many decades now, the construction industry has been encouraging and using precast beams, floor and wall slabs, stair flights, and other elements, and recently even small precast building components like toilets and closets, to reduce hazardous site work at height.

Two interesting developments offer alternatives to scaffold and formwork for high-rise construction, as follows.

(a) Prefabricated, prefinished, volumetric construction

Abbreviated to 'PPVC', this involves casting and curing complete segments of high-rise units or even entire units at floor level in a factory, outfitting them with all the utilities and devices required in the finished structure, transporting them to the building site, and placing them by one or more cranes one on top of another, to complete the planned high rise.

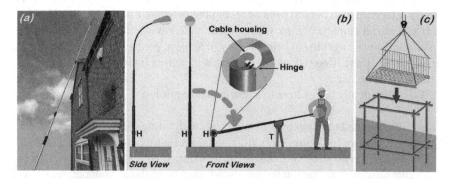

FIGURE 3.1   Fall avoidance examples.

(a) Telescoping long-handled water jet. [Source: Own work, considerably modified from the original image. Author: Unknown. Link: www.domesticcleaningpoles.co.uk/product-page/gutter-cleaning-pole-tool-gutter-cleaning-brush]

(b) Lamp post hinged for bulb changing. [Source and Author: Own work.]

(c) Guardrail integrated with work platform. [Source and Author: Own work.]

As the horizontal and vertical connections between different units can be made from inside the structure accessed by the permanent staircases forming part of the final plan, at the respective floor levels, there is no work at height involved.

The technique was first demonstrated it as 'Habitat-67', at the Canadian Pavilion of World Exposition in 1967 at Montreal by the inventor of this technique, Moshe Safdie [Ref. **3.1**]. Figure 3.2(a) shows a close-up view of Habitat-67, and Figure 3.2(b) a view of a PPVC unit in Malaysia being lifted into place by crane in 2021.

It is to be noted that the lifting is done at the four corners of the unit, not being attached directly to the crane hook, but to the four corners (and along sides as needed) of a steel frame of the same plan dimensions as the unit, the frame being attached to the crane hook. Otherwise, the inclined lifting cables from a single crane hook would introduce large lateral compression forces into the upper layer of the unit, designing for which would be extra cost.

In the pioneering effort of 1967, inventor Moshe Safdie constructed 354 identical and completely prefabricated modules stacked in various combinations and connected by steel cables. The apartments varied in shape and size and were arranged in different configurations to give buyers and tenants a wide choice.

It took a few decades for the technology to develop to a cost-effective method of construction, and many countries (including Singapore and India) have taken it up as a safer and more feasible alternative to in situ concrete construction.

It should be noted that while the use of precast elements and PPVC has eliminated fall hazards from the worksites, many of the fall hazards for single-storey construction have been moved into factories. The advantage, however, is that it is easier to

FIGURE 3.2   Fall elimination examples.

(a) Moshe Safdie's Habitat '67. [Source: Own work. Author: Bohemian Baltimore. Link: https://commons.wikimedia.org/wiki/File:Habitat_67_Montreal_Quebec_10.jpg]

(b) Lifting of PPVC unit. [Source: Open Access paper, "Comparative Study between Prefabricated Prefinished Volumetric Construction (PPVC) and IBS 2D: A Case Study of School Extension Project in Malaysia", IOP Conf. Ser.: Earth Environ. Sci. 1022 012016. Authors: Tambichik, M.A., Z S Sherliza, and N A Abdullah. Link: https://iopscience.iop.org/article/10.1088/1755-1315/1022/1/012016/pdf]

(c) 3-D Printed House. Screenshot from video by Alfredo Milano, of house designed by Mario Cucinella Architects and built by World's Advanced Saving Project (WASP) Company of Italy in 2021. Link: https://commons.wikimedia.org/wiki/File:Eco-sustainable_3D_printed_house_%22Tecla%22.jpg]

manage these fall hazards inside the controlled environment of the factory than out in the open, and most of the work will be at ground or floor level.

(b) 3-D Printing

Imagine your inkjet printer nozzle enlarged thousands of times, to about 3 cm or more, and instead of inks for the three primary colours red, blue, and green plus black, you feed fine aggregate concrete with extremely quick-setting and hardening properties [Ref. **3.2**].

Then, we have a 3-D concrete printer, which, guided by appropriate computer software such as building information modelling (BIM), can print a structure of almost any desired shape and size! Figure 3.2(c) depicts the schematic for 3-D printing of a building.

This technique may be used for any shape, size, and height of structure. The main hurdle for this technology is the material that will dry and harden within a matter of minutes but which will be as strong as traditional construction to last for a long time.

In experimental mode for the last two decades, 3-D printing has become more accessible and attractive to the construction industry, with even developing countries like India [Ref. **3.3**], pushing the technology for fast implementation.

3.2.3 PREVENTION OF ACCESS

Heads of families at home and authorities at worksites often resort to preventing access in situations where individuals or groups would be exposed to a common danger.

Common examples are:

*   Baby gates for stairs, to keep babies from climbing upstairs;
*   Schools being declared closed when a storm is imminent or raging;
*   White-collar professionals being asked to work from home during the onslaught of the COVID virus;
*   Public being banned from entering a radioactive, fire, or other hazardous zone.

These would be similar to fencing in or leashing out pets from running amok.

3.3   EDGE PROTECTION, GENERAL

3.3.1 COMMON CURRENT PRACTICE

By far the easiest, simplest, and cheapest method to prevent falls for all who wish or need to access and work at heights is the barrier against falling off a certain area. This logical solution is termed as 'edge protection' – by means of a guardrail or parapet (also called 'balustrade' topped by a 'banister').

This has been done for centuries – although as pointed out in Chapter 2, our ancestors might have omitted such edge protection, when and where they did not feel the need to go to the expense and trouble for it – or just did not think of it! – and instead adapted themselves to the risk intuitively.

Babies are traditionally kept inside cribs by side guards. Beds for senior citizens and hospital patients are protected from falls by side rails. Almost all stairs, balconies, bridges, and public places at heights today have edge protection.

In normal adult world, when a barrier is erected to prevent a person going over an edge, we call it 'guardrail' if temporary or 'handrail' or 'parapet' if permanent. When it is to prevent some person or animal running away from or entering a certain zone – or accessing a dangerous piece of equipment or material, we call it as 'fencing'. Parapets for permanent structures are usually somewhat higher than for temporary structures because the former will be used by people of varying ages and behaviours, while the latter by trained and monitored personnel.

Sometimes, a different connotation is attributed to the terms: A 'guardrail' may refer to fall prevention along open edges at height, and a 'handrail' to rails to assist in climbing up or down stairs.

Edge protection for temporary structures such as scaffolds requires separate regulations and specifications.

3.3.2   HEIGHT OF EDGE PROTECTION

The height of edge protection varies between countries, and in the same country, depending on purpose and location.

For instance, for permanent structures occupied by the general public, the parapets are higher than for structures used by workers in construction sites or factories.

More often than not, guardrail height appears (logically) to depend on the height of the users. From a mini-research project in connection with an accident investigation [Ref. **3.4**], I discovered that (within the limitations of that study) the optimum height of protection to prevent a worker falling while leaning on or over the guardrail should be equal to or more than the height of his centroid, namely about 0.55 times his height.

The world average height of males of about 1.70 m translates to a guardrail height of 0.94 m, conveniently rounded to 1 m, which is the most common height used!

In the USA, the recommended parapet height is 42 in (1.07 m), as shown in Figure 3.3(a); it has been followed from 1897. This is applicable to workplace edge protection in particular.

In Denmark and Netherlands, it is 1.2 m (47 in.). In Nepal, it is 1.0 m (39 in.) and in India it is between 1.15 and 1.3 m (45 and 51 in.). In Singapore, regulations for parapets in Housing & Development Board flats specify 1.2 m (47 in.), and for guardrails in workplaces 1 m.

The parapet height varies also with the slope of the surfaces it guards. The International Building Code [Ref. **3.5**] recommends parapet heights at the edges of sloping roofs, from 42 in (1.07 m) for slope of 1 vertical in 4 horizontal to 54 in. (1.37 m) for slope of 1 in 2.

In some countries, the height of handrail above the 'nose' of a step for stairs is shorter than the guardrails at horizontal plane edges by about 3–6 in (7.5–15 cm).

The American practice is shown in Figure 3.3(b). In India, the stair handrail is set at 90 cm (35 in.). In Singapore, the handrail height for stairs was recently revised from 90 cm (35 in.) to the standard height for work platforms, namely 100 cm (39 in.).

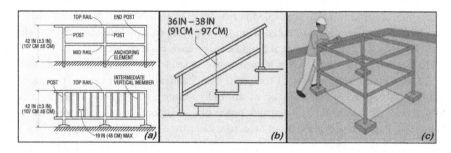

FIGURE 3.3   Edge protection requirements.

(a) Guardrail types and dimensions. [Source: OSHA Standard Number: 1910.29, Fall protection systems and falling object protection – criteria and practices, Figure D-11. Link: www.osha.gov/laws-regs/regulations/standardnumber/1910/1910.29]

(b) Handrail requirements. [ibid., Figure D-13.]

(c) Guardrail around opening. [Source: OSHA Technical Manual (OTM) Section V: Chapter 4, Figure 2. Link: www.osha.gov/otm/section-5-construction-operations/chapter-4]

In this book, as far as edge protection height is concerned, we will treat all open edges, horizontal or sloping, the same from a 'holistic' viewpoint.

3.3.3 OTHER EDGE PROTECTION REQUIREMENTS

All edge protection intended to prevent people from falling over the edge must be strong and sturdy to bear any vertical and lateral loads that may be applied to it by users. Often twice an average adult's weight should suffice as the design force.

If at any time any larger lateral load is anticipated to be applied on a parapet – as from the hook of a suspended scaffold for instance, or for steeper roof slopes – it must be designed for this extra force also.

Typical guardrails will have two or more rails with the top rail at the 'standard' height for that country and facility. To prevent accidental slip through of a person through open grid barriers, the rails must not be spaced more than about 50 cm (20 in.) or less if children are expected to be around.

Also, horizontal intermediate railings must be avoided as children (and adults too) will be tempted to climb on them to reach higher or see better.

Guardrails must also be provided to all other locations where there is an opening on the inhabited work area and which have the potential for a fall. Common examples are openings on the floors of attics or work platforms reached by a ladder, glass-covered skylights openings on terraces, etc. [see Figure 3.3(c)].

As an alternative, openings may be covered by a fixed or movable cover, strong enough to bear the weight of a person safely; some indication that the cover is over a hole will also help.

It goes without saying that handrails, which (unlike parapets and guardrails which are intended to prevent falls) are used more to help people walk along corridors or climb up and down steps and must be smooth so as not to injure the skin on the user's palm sliding along it.

Most important of all, the handrail must be small enough and of the right shape to be actually held within the hand. Many regulations even recommend the addition of a second handrail of smaller diameter at a lower height, just for the sake of shorter people and children.

Unfortunately, the so-called aesthetics of modern culture frequently ignore safety criteria or overcome practical concerns in edge protection. To blame the architects and interior designers who suggest such designs or the clients and owners who want them or accept these designs is not relevant, but the consequences surely can be literally deadly.

In some 'posh' homes of fashionably rich folks, stairs do not have any handrails or other barrier. We can only hope the owners of these modern homes do not have in their family – and do not invite to their parties a family with – a toddler, a pregnant lady, or an older person!

3.3.4 CASE STUDY ON EDGE PROTECTION

Figure 3.4 depicts a case study in my own experience.

I found a flight of steps with a handrail in a children's park close to my apartment Figure 3.4(a), which was so large – Figure 3.4(b) – I called it a 'hug-rail'! I sent its photo to the concerned local authorities, describing the fall risk it posed, and also proposing my solution for a smaller handrail and vertical bars to be welded to the existing handrail.

The township council went much farther than I had suggested and, in a few months, replaced it with a conventional stainless-steel handrail [see Figure 3.4(c)].

More details of the case study may be found in my two books, Refs. **3.6** and **3.7**.

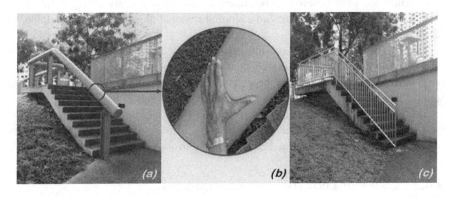

FIGURE 3.4 Case Study of edge protection.

Source and Author: Own work.
(a) Steps in a children's park.
(b) Close-up of the handrail.
(c) Replacement handrail.

3.3.5 WHEN EDGE PROTECTION MUST BE RAISED

The principle used in the definition of edge protection height for horizontal walking surfaces must be extended to all working surfaces. Thus, for ramps, or where a roof or corridor has an activity area higher than the rest of the space, the erected barrier must also be raised with it, so that when a user accesses the higher levels, he is still protected to the desired height,

This topic will be dealt with in more detail later.

3.4 EDGE PROTECTION FOR WORK PLATFORMS

Collective edge protection by means of standard guardrails is the norm at workplaces around the world.

3.4.1 WORK PLATFORM

Although a platform must obviously be strong and sufficiently rigid not to deform too much or sway, regulations prescribe very strict requirements for platforms in temporary structures like scaffolds and concrete formwork, mostly because components of temporary structures will be erected, used, and dismantled at different workplaces repeatedly until they deteriorate below required standards.

The integrity of the platform will be critical to fall safety measures.

The platform must be from materials which will not deteriorate too soon with repeated erection, use, and dismantling, or exposure to extreme weather conditions.

The factor of safety against failure varies from 3 or more on the maximum anticipated worst combination of service loads during its expected life span.

There must be no gaps between the planks or plates composing the platform, beyond a small allowance of a few millimetres (fraction of inch).

Overlaps too must be avoided because they would pose a stumbling ('trip') risk to workers. Some codes recommend a sloping insert at an overlap if unavoidable, to facilitate safe traversing.

The ends of the planks or plates must not extend beyond supports beyond a certain limit, usually four times their thickness. To prevent their dislodgment from the supports during use, the boards must project a minimum length (about 5 cm) beyond the support and be tied down or otherwise anchored to the tops of the supports.

Safe access and egress to the work platform from lower levels should be provided by means of stairs (with guardrails) within the scaffold frame or by vertical ladders on the outside (or steeply inclined ladders on the inside) attached to the frame.

3.4.2 EDGE PROTECTION FOR SUPPORTED SCAFFOLDS

For scaffolds supported on the ground or other base, for concrete formwork and other temporary structures at the workplace, most countries have specifications for standard guardrails, as in Figure 3.5(a), which happens to be the one recommended in Singapore. Instead of rails, weld mesh may be used, as in Figure 3.5(b).

The toeboard of height 90 mm (3.5 in.) or more is to prevent tools or debris falling over the edge and harming a pedestrian or worker below.

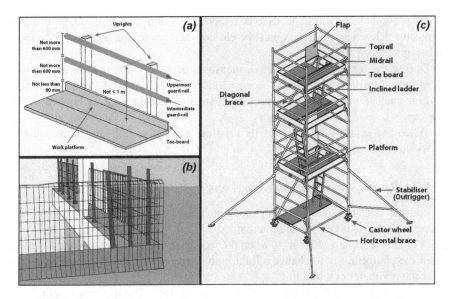

FIGURE 3.5 Edge protection.

Source: Code of Practice for Working Safely at Heights – Second Revision 2013. Author: Singapore Workplace Safety and Health Council. With permission from the author.
(a) Typical guardrail. [Source Figure 6.9.]
(b) Mesh barrier. [Source Figure 6.1.]
(c) Guardrail on mobile scaffold. [Source Figure 6.10.]

On a wet platform, a person too may slip out under the guardrail, and thetoeboard would save him from falling. Some may add a net or mesh fabric on the rails for extra safety of falling smaller debris, tools etc. – even small-framed persons.

America considers cross-bracing of a scaffold as the equivalent of one of the two horizontal rails, depending on whether the crossing of the two inclined braces is at the recommended level of the top rail or mid-rail.

The other horizontal rail must be inserted to complete the safety requirement that no vertical gap may be more than (about) 50 cm (or 20 in.)

Climbing to the platform on the bracings or platform side supports is strictly forbidden in most countries.

Figure 3.5(c) depicts a mobile scaffold provided with guardrail edge protection. The figure also illustrates a few other fall safety features:

(a) At the point of access to the platform by stairs or ladder, there should be a swing gate in the handrailing if access is from outside or a hinged flap for the horizontal opening in the platform if from the inside – the latter marked F in the figure. The gate or flap must remain closed except during entry or exit.

(b) Many codes' have limits on ratio of height to lesser of the two plan dimensions, usually 3 or 4. When the height-to-width ratio exceeds occasionally by not more than 100%, the deficit in base width occasioned by the increased height may be made up symmetrically with 'stabilisers' or 'outriggers'.

If a portion of a guardrail needs to be temporarily omitted or removed, for instance to receive deliveries, such gaps must be closed when not in use, by means of a movable segment or gate.

During the temporary gap period, measures must be taken to restrain the user from falling through the gap. If the guardrail has to be removed frequently during certain times, a temporary edge protection such as a chain must be hooked across the gap.

When the guardrail is temporarily missing, alternative fall safeguards must be provided to maintain 100% tie-off.

3.4.3 Raising Guardrails in Workplaces

At construction sites, factories, shipyards, etc., raising of guardrails to keep up with rising work levels is critical and mandatory.

A common scenario for this is when the worker is briefly tasked beyond his reach from the platform, say to change a light bulb or run some conduit along the ceiling near the perimeter of a building.

Then, he needs to climb on something to do his task. The moment he climbs on a ladder or a stool placed on the work platform, he violates the logic that the edge protection must be the regulation height above his foot level.

In such situations, as and when the person is required (as at work) or allowed (as in tourist attractions) to go higher than the prevalent edge-protected level, the occupier must also ensure edge protection for the standard height above this higher level.

The simplest and fastest way to provide the required extra height of edge protection would be to use a stepladder of required height with its own guardrail L and chain guard C at top, as shown in Figure 3.6(a).

Needless to say, while the worker is using the ladder, the ladder must be tethered to the lower-level guardrail to prevent tipping over, any wheels must be locked, and continuous supervision also must be provided. Even so, the user must be extra careful not to lean too far from his feet position.

An alternative, as shown in Figure 3.6(b), is to extend the existing guardrail temporarily to suit the height by which the user climbs from the platform. This could be done by attaching vertical tubes (unshaded, E in figure) to the existing posts (shaded, G in figure) by means of couplers and adding horizontal rails.

Commercially, guardrails are available which have built-in telescoping extension capabilities as shown in the inset of the same figure, at some greater cost.

Another interesting situation arises when temporary edge protection for different heights of tasks is needed occasionally, as when servicing a mobile tanker.

A fast and economical solution is a mobile scaffold with telescoping steps and handrail, and a guardrail enclosure attached at its top, as shown in Figure 3.6(c).

The unit is moved to the job, its horizontal location reached by the wheels, and the vertical access level adjusted by the telescoping ladder; then, the worker simply climbs to the top and works on the task, from within the enclosure, as at A in the figure.

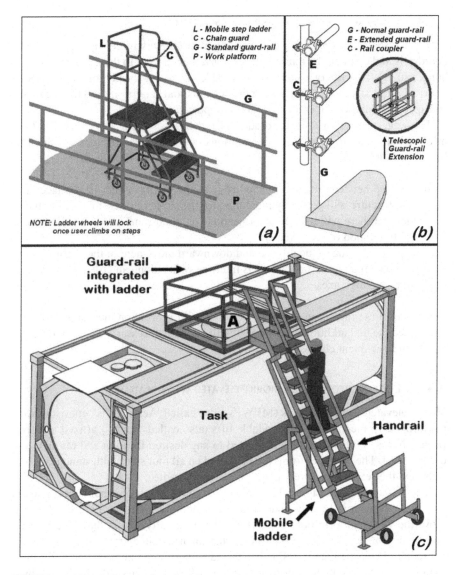

FIGURE 3.6   Vertical extension of edge protection.

(a) Use of stepladder. [Source and Author: Own work; ladder figure taken from public domain.]

(b) Pipe and coupler extension. [Source and Author: Own work.]

(c) Mobile ladder with guardrail. [Source: Code of Practice for Working Safely at Heights – Second Revision 2013. Author: Singapore Workplace Safety and Health Council. With permission from the author.]

3.4.4 EDGE PROTECTION FOR SUSPENDED SCAFFOLDS

Suspended scaffolds, consisting of a working platform with three edges protected by guardrails and open on the fourth side for convenient access to the assigned task, are hung, usually from the two short sides by cables from winches on the roof (or other floor) of a tall structure. They face more hazards and require more fall safeguards and special design considerations than do supported scaffolds.

The edge protection criteria to address fall hazards from suspended scaffolds must include the following two considerations both of which will increase the likelihood of falls:

(a) The gap between the unprotected edge of the platform and the job face of the structure will generally more than for the supported scaffold, due to the fact that there will usually be projections on the façade, and this will make the likelihood of falls greater;

(b) During use, due to the upward and downward movement of the scaffold, as well as wind gusts, the scaffold is likely to swing along and tilt about any or all of its three axes.

Because of these increased fall risks, collective fall prevention measures will cease to be effective and additional fall control will become necessary. These will be discussed in subsequent chapters.

3.4.5 EDGE PROTECTION FOR MOBILE ELEVATED WORK PLATFORMS

Mobile elevated work platforms (MEWPs), also called 'Aerial Lifts', consist of work platforms atop mechanically extendable lifts (also called 'hoists') affixed on motorised vehicles, so that they may be moved to any desired location and raised to any desired level. The work platforms are protected on all four sides with standard guardrails, and the set-up is referred to as a 'basket' or 'cradle'.

The scissor lift [see Figure 3.7(a)] will only move vertically and, hence, must be moved to a location directly under the task. The boom lift [see Figure 3.7(b)] can move the work platform both vertically and horizontally within half a hemisphere and hence is more useful; the type shown has an articulated arm, but it may also be telescopic. The articulated version has the advantage it can be bent into a 'C' shape to move the platform under a bridge from a vehicle positioned on the bridge roadway, for the platform occupant to inspect or work below the bridge.

The circular inset in Figure 3.7(c) presents fall safeguard of another type, namely a lanyard connecting the D-ring (or waist belt) of the user to an anchor ring at the base of the basket highlighted by a smaller circle; a similar ring is provided at the other three corners.

This fall restraint – to be discussed in the next chapter – will prevent the user from falling if the worker leans beyond the guardrails to a limited extent or to avoid his falling off to any sudden instability of the basket.

The single-mast tower in Figure 3.7(d), also with guardrails on all four sides of the platform, is also an MEWP, acting like a cargo lift, transporting workers and essential materials to and stationed at any desired height for work on the facade of the

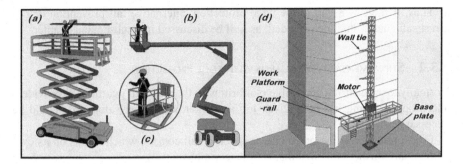

FIGURE 3.7   Mobile elevated work platforms.

[Source: Code of Practice for Working Safely at Heights – Second Revision 2013. Author: Singapore Workplace Safety and Health Council. With permission from the author.]
(a) Scissor lift. [Source image, Figure 6.11.]
(b) Articulated boom lift. [Source image, Figure 6.11.]
(c) Anchor links for fall restraint. [Source image, Figure 6.12.]
(d) Mast tower. [Source image, Figure 6.14.]

completed building; it may also be used for entry into the building at any opening, for interior work. The tower is tied to the structure at intervals for stability. Twin-tower models are used for longer facades.

MEWPs are very efficient and cost-effective for reaching and working at points at height safely, without the effort and time needed to erect supported or suspended scaffolds, or where such scaffolds are infeasible. If entry from it into another zone is intended, the facing guardrail must have a gate, and safe provision – such as a foot-bridge with guardrails – must be made to cross the gap between the basket and the intended zone.

3.5   ADDITIONAL CONTROLS WITH EDGE PROTECTION

3.5.1   NEED FOR 100% FALL MANAGEMENT

A question may arise on what to do if there is a gap in the edge protection. This may happen during erection or in normal use (as for delivery of materials or equipment). Do we just caution the workers to be extra careful while negotiating the gap in the edge protection, and put up warning signs as reminders?

Not so. The logic, and hence the regulation on it, are very clear: Edge protection has to exist for the entire length of work at height, for the full duration of the work – in short, the fall prevention (or protection) measures shall be available with '100% Tie-Off'; this will be discussed in the next chapter.

It is unrealistic to believe that a 'small' gap or a 'short' time without edge protection should be acceptable because the workers are all responsible adults, with their own safety and health at stake. Accidents do not take too much space or too long a time.

Any gap wider than the maximum allowable gap between the platform and the job face (normally 300 mm) is an invitation to a fall accident.

Thus, if there is a gap in the edge protection, alternative fall prevention or fall arrest safeguards shall be provided, as will be discussed in a subsequent chapter.

3.5.2 SUPPLEMENTARY SAFEGUARDS WITH EDGE PROTECTION

Normally, once a strong and stable platform with the standard guardrail is provided, and proper training has been given to the worker, no further safeguard should be necessary.

Edge protection is the one unique collective fall control, which stands on its own without any further embellishments or reservations.

Hence, there is no need for workers on properly designed and erected scaffold platforms conforming to the regulations to be given additional work restraints or fall arrest PPE.

However, supplementary safeguards may be provided in the form of the following to support or even supplant edge protection:

(a) Warning signs, by means of black/yellow or red/white striped tapes or word signboards, informing the workers of the presence of the barrier, and the hazard addressed, such as 'Warning: Hole below!"

(b) In certain countries like the USA, a red stripe on the floor or a striped tape at waist height and located sufficiently behind the unprotected edge would serve to define the limit beyond which the worker shall not stray – placing the responsibility for fall safety fully on the worker who normally takes it as part of his duties.

(c) Infrared light or radio frequency beams may be set up at strategic locations to trigger warning signals to workers if and when they get too close to an unprotected edge.

REFERENCES FOR CHAPTER 3

3.1 Merin, G., "Architecture classics: Habitat 67/Safdie architects", *ArchDaily*, 9 February 2023. Retrieved on 28 June 2023 from: www.archdaily.com/404803/ad-classics-habitat-67-moshe-safdie, ISSN: 0719-8884

3.2 Amelia, H., "How does a concrete 3D printer work?", *3D Natives*, 8 January, 2021. Retrieved on 28 June 2023 from: www.3dnatives.com/en/how-does-a-concrete-3d-printer-work-080120215/#!

3.3 Anand, N., "L&T is in talks with builders to popularise 3D concrete printing technology", *The Hindu*, 12 September 2022. Retrieved on 14 February 2023 from: www.thehindu.com/business/Industry/lt-is-in-talks-with-builders-to-popularise-3d-concrete-printing-technology/article65882793.ece

3.4 Krishnamurthy, N., "Worker fall from mobile scaffold", *International Journal Forensic Engineering*, 1(1): pp. 21–46, 2012.

3.5 ———. *International Building Code 2018*. International Code Council (Headquarters), Washington, DC, USA.

3.6 Krishnamurthy, N., *Introduction to Enterprise Risk Management*. Published by author through Amazon, USA, pp. 57–58, 227–229, 2016; also, Partridge Publishing Co., Singapore/USA. ISBN: 978-1-543-75472-8.

3.7 Krishnamurthy, N., *Essays in Forensic Engineering*. Published by author through Amazon, USA, 238 pp, 2017; also, Notions Press, India, 215 pp, 2018. ISBN: 978–1–644–29656–1.

4 Individual fall prevention

4.1 ASPECTS OF INDIVIDUAL FALL PREVENTION

4.1.1 SITUATIONS FOR INDIVIDUAL FALL PREVENTION

Collective prevention of falls may not be feasible in certain situations, as for instance:

- For individuals who need support while standing, walking, etc., as otherwise they are liable to fall and injure themselves;
- When edge protection is difficult, impossible, expensive, time-consuming, or inconvenient to erect, or is required only for such short times and/or for such short lengths that erection of a barrier may not be time- or cost-effective;
- While the collective edge protection is being erected or dismantled.

There are many situations where vertical distances must be negotiated, for construction, maintenance, rescue, or other purposes, in which there is no question of collective prevention of falls applicable to horizontal or slightly sloping surfaces.

In such cases, the next logical choice is individual prevention, that is, preventing individuals (a) from falling when and after they access heights, and also while on level, and (b) from dropping off into a free fall while negotiating a height on purpose.

4.1.2 SCOPE OF INDIVIDUAL FALL PREVENTION

More individuals fall and get injured outside the workplace than inside, even at level, let alone at heights. Apart from offering some general guidelines and platitudes, falls of the general public cannot be collectively prevented, particularly in their homes and personal haunts. Hence, individual fall prevention is a significant need and deserves proper attention.

Individual fall prevention is lower in control hierarchy than collective fall prevention, as it will need assessment of the likelihood and severity of fall hazards that each individual faces, and his capabilities to adopt the guidelines and safeguards recommended, followed by training, guidance, and monitoring of that individual before and while he is using the prevention methods and devices.

It also needs the individual's understanding of the hazards as well as of the preventive safeguards proposed, and his acceptance and correct implementation of the safeguards.

Individual fall prevention devices fall under the category of PPE, which is accepted worldwide as the least effective and last of safety controls in the fivefold hierarchy, after Elimination, Substitution, Engineering Controls, and Administrative Controls.

Thus, they should not be adopted before hierarchically higher options are considered and found infeasible.

Fortunately, fall restraint devices constituting the PPE are not unduly complicated or uncomfortable to wear and use, nor are they too expensive. In fact, with just a little bit of training and minimal supervision, individuals have been using such PPE as walking aids with no problem for centuries.

Individual fall prevention may be categorised in two ways for our purpose:

(i) 'Fall restraint', 'Work restraint', or 'Travel restraint' to prevent falls while the user moves to or on a horizontal or slightly – up to say 1 vertical in 10 horizontal – sloping plane, while his weight is borne by the surface below his feet;

(ii) 'Rope access' or 'Work positioning' to prevent falling while the user moves or is stationed vertically in air or on a steeply sloping surface, while gravity is acting to pull him downward.

4.2 GENERAL FALL RESTRAINT

The most common method to prevent individual falls is by means of supporting the person intending to move in a horizontal or nearly horizontal plane, in a way that he will not (a) collapse while standing or walking or (b) reach past the unprotected boundary of the safe (non-fall) zone at height.

(We will skip the walking stick as too basic to include here – except to point out that modern medical versions have adjustable lengths and multiple toes for better support!)

4.2.1 FALL PREVENTION AT LEVEL

While we cannot realistically control the general public from level fall hazards, we can and do plan to support children, old people, and sick or injured patients who under rehabilitation to normal mobility, from falling while standing or walking on level. Some version of a 'walker' is in common use for such individuals of all ages who are learning or re-learning to walk and, due to health problems or old age, tend to lose their balance while standing or walking.

Figure 4.1 shows various devices used for this purpose.

(a) A child's walker as a safe self-training device to learn walking – the pelvic support prevents body collapse;

(b) Adult under rehabilitation training, very similar to (a), supported by a pelvic harness, enabling him to walk around without risk of falling; the pelvic harness is also known as a 'sit harness' or 'seat harness', because if the user gets tired or collapses, he simply sits on the bottom strap.

(c) A suspension walker frame with chest harness and pelvic support to enable the user to walk around without fall risk;

(d) A standard walker used by many senior citizens and most recovering patients to avoid falling due to loss of balance.
   Except (d), where the user lifts the walker to wherever he wishes to step next, walkers are usable only on level ground or floor, or on gentle slopes as for wheelchair ramps.

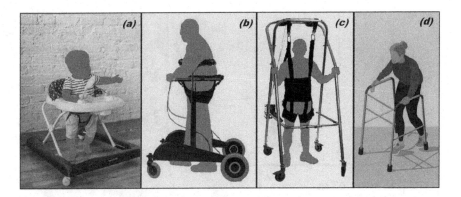

FIGURE 4.1 Walker support devices.

Source: Modified or redrawn from public domain sites such as Pexel, Freepik, etc.
(a) Baby walker.
(b) Pelvic harness.
(c) Walking suspension frame.
(d) Adult walker, with or without wheels.

4.2.2 FALL PREVENTION FROM HEIGHTS

Unlike fall-prone persons mentioned in Section 4.2.1, able-bodied persons likely to stray to unprotected edges or into dangerous areas, when edge protection is infeasible, will need their movements to be restricted to within a permissible or safe zone, beyond which they would face – or create – problems. This is best done by a leash, also named (more politely!) a tether.

Among the general public, the most common example of the use of leashes is to control the movement of pets [see Figure 4.2(a)] and (somewhat controversially) hyperactive children. Although this has nothing to do with heights, the strategy becomes a very good choice to prevent falls to persons moving close to unprotected edges on level or on slight slopes.

4.3 FALL RESTRAINT AT THE WORKPLACE

4.3.1 FALL RESTRAINT

Considerations for individual fall prevention at work sites are different from those discussed above for the public, as the individuals at the workplace will be able-bodied and trained for the task, and the chief concern here will be falls from height, more than falling on level or getting into trouble.

Figure 4.2(b) illustrates the principle of fall restraint, namely, to give the user a lanyard connecting his waist belt to an anchor, the lanyard length not more than (– actually about 15 to 20 cm shorter than) the distance from the anchor to the unprotected edge P, so that there is no likelihood of his falling.

The point to remember is that the restraint user should not and will not fall!

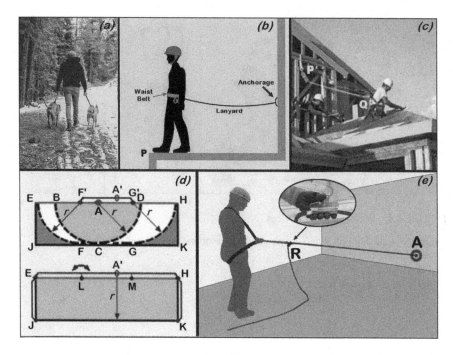

FIGURE 4.2 Travel and fall restraints.

(a) Dogs on leash. [Source: Unsplash. Author: Photo by Meritt Thomas. Link: https://unsplash.com/@merittthomas]
(b) Work restraint. (P – Unprotected edge. [Source: Code of Practice for Working Safely at Height, Figure 8.2. Author: With permission from Singapore Workplace Safety and Health Council. Link: www.tal.sg/wshc/resources/publications/codes-of-practice/code-of-practice-for-working-safely-at-heights-english]
(c) Work restraint on sloping roof. [Source: Fall protection in residential construction. Author: OSHA. Link: www.osha.gov/residential-fall-protection/guidance]
(d) Enlarging work zone with fall restraint. [Source and Author: Own work.]
   Upper figure: Use of sliding bar; Lower figure: Combination of fall restraint and edge protection.
(e) Rope-grab for adjusting length of work restraint. [Source and Author: Own work, inset rope-grab generic item.]

Apart from lanyard length being a little less than the distance to the unprotected edge, other requirements of a fall restraint are as follows:

- The anchor is best at or close to his waist level. The lanyard may be connected to the D-ring of a chest harness instead of the waist belt, with the anchor correspondingly higher, if possible. Figure 4.2(c) displays a worker with a fall restraint PQ.
- The anchor may be higher or lower than the connection point to the user by an amount V, as long as the horizontal distance H, computed as $\sqrt{(L^2-V^2)}$, is less than the distance to the unprotected edge. Anchor at higher level is

better than at lower level, because in the latter case, the person's squatting down may increase the horizontal mobility to the point of slipping over the edge, although not freely falling. In any case, the anchor should not be lower than the foot level.

- When the lanyard is inclined due to anchor not being at the same level as the waist belt, to avoid the likelihood of the waist belt slipping up or down, the worker may wear a chest harness or even a combined chest and pelvic harness, this last combination approximating to a full-body harness but without the shock absorber (which will be described later).
- The anchor strength needs to be only two to three times the maximum weight of the user including tools.

4.3.2 ENLARGING THE AREA COVERED BY FALL RESTRAINT

There is one inconvenience with the work restraint: The lanyard of length r attached to a single stationery anchor will allow the user to only work within the full or part of a circle of radius r! As generally a work platform is adjacent to the task, only a semi-circular zone is available to work, such as BCD in the upper part of Figure 4.2(d), with the anchor at A.

This is not really a problem. We can expand the work zone by providing a ring at the end of the lanyard sliding along a horizontal bar or lifeline, instead of being fixed at one location. The extra critical consideration is that where the increased radius goes past any portion of the unprotected edge, edge protection must be erected to prevent a fall over that portion – certainly not stop by warning the user to 'be careful'!

Thus, as shown in Figure 4.2(d), by providing a slide rod over the length F'G', the user can access the expanded area BFGH of the rectangle F'FGG' of width r and the two quadrants EF and GH of radius r. Needless to say, neither F'E nor G'H can be more than r, for the user to remain on the platform.

Even now, the corners J and K are not accessed.

If the entire rectangle EJKH needs to be accessed, then we must provide a slide rod or cable for the full length EH of the rectangle, with the same lanyard of length r, as shown in the lower part of Figure 4.2(d).

However, now, he will be exposed to fall risk along the short edges JE and KH. Therefore, along these two short edges EJ and HK, we must provide standard guard rails. It may still be cheaper and faster than erecting guardrails over the entire perimeter EJKH, especially if it is for a short duration.

If there is an intermediate post on the slide rod or cable as at L and M, they must be negotiated with 100% tie-off as explained in the next section.

If the area is not narrow, the anchor may be set up at any convenient spot in the middle, and similar analysis conducted by drawing circles of various required radii to reach the task areas, and wherever the circle cuts the boundary, erects edge protection.

This requirement to provide edge protection where there is a deficiency in individual fall prevention by fall restraint is the exact reverse of what we have seen in the previous chapter, namely the need to provide fall restraint (or other alternative safeguards discussed in subsequent chapters) when there is a gap in edge protection.

4.3.3 ROPE GRABS FOR VARIABLE LENGTH FALL RESTRAINT

A 'rope grab' is a device which when clamped around a lanyard ('rope') grabs it tight, letting go only if its clutch handle is pressed. It should be evident that if a rope grab is attached to the end of the user's short lanyard from the frontal 'D' ring (so called for its shape) of his belt or harness, and clamped around the cable from the anchor, then the user can release the grip by pressing the clutch handle and adjust it to whatever length he wants at any particular time. In some models, the grab is normally free to move, and the user must press the handle to lock-grab it; however, this is not as fail-safe as the former version because a careless and loose lock would let the user slide into the fall. In many models, sudden jerks on the rope will lock the device.

The use of such a rope grab device for variable length fall restraint is illustrated in Figure 4.2(e), with the rope grab R shown enlarged in the oval inset; A is the anchor. A nice video of its use is available in Ref. **4.1**.

No doubt this will give a lot of freedom to the user and less installation effort to the employer. But the entire responsibility for safety will rest squarely on the user's own shoulders, also requiring additional training and extra supervision for legal coverage by the employer.

Actually, it works quite well in the USA and other countries where the work force is reasonably homogeneous and shares responsibility for safety with the management, but not so well elsewhere.

4.4 100% TIE-OFF

4.4.1 NEED AND PROCEDURE

Technically and professionally, the idea of continuous protection is logical. Unfortunately, it is human nature – and quite tempting to employee and employer alike – to assume that a relatively small gap in the edge protection or an obstacle can be negotiated without much fuss.

This is not only illegal but also unrealistic, because most accidents happen in situations that people believe to be simple and safe, suddenly and fast.

As has been mentioned earlier, the ideal fall control is one that provides fall prevention (or, if prevention is infeasible, fall arrest) safeguards over 100% of the area of activity, for 100% of the time, without imposing any further guidelines or requirements on the user. This is called '100% Tie-off'.

Fall restraint is the best answer of how to negotiate gaps or obstacles with 100% tie-off, as it will preserve the intent and planning of fall prevention, although individually instead of collectively, for a few small parts of overall safety.

To accomplish this, the user must be given a 'Double lanyard' or 'Twin-tail lanyard', which is a pair of cables to be attached from the same waist belt or harness D-ring to two different anchors, as shown in Figure 4.3(a).

Let us say there is a gap in edge protection or an obstacle at L on a work platform, and the user, now at S, wants to move to T past the gap or obstacle.

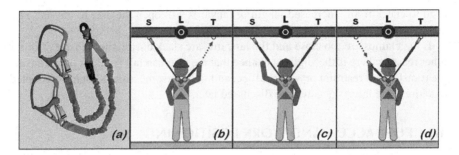

FIGURE 4.3   100% tie-off.

(a) Twin-tail or double lanyard. [Source and author: Generic item.]
(b, c, d) Steps in transferring restraint from S to T. [Source: Generic item. Author: Modification, own work.]

The procedure is quite simple, although workers may need repeated demonstrations and explanations of the logic and safety significance of each step, before they are able to use it with confidence.

(a)   User is currently attached to the left segment at anchor S [see Figure 4.3(b)].
(b)   Retaining the attachment at S, he takes the other lanyard from inside his belt (or from its pouch).
(c)   He attaches this second lanyard to the anchor at T on the next segment at right [see Figure 4.3(c)]. Both lanyards are attached now.
(d)   He can now remove the attachment at S and tuck it into his belt (or its pouch) [see Figure 4.3(d)].

He is now safely attached at anchor T and may proceed with his task in this next segment, having negotiated the interruption in edge protection with 100% fall prevention.

4.4.2   Applications of 100% tie-off

The technique may be adopted in the fall restraint sliding anchor mentioned in the previous section, where it meets with an obstruction such as a support post, as at L and M in Figure 4.2(d).

With proper training and appropriate self-discipline and confidence, this twin-tail lanyard technique for 100% tie-off is being widely used to climb the hundreds of rungs of the vertical ladders for wind turbines or electrical towers, and for steel erection with just rings welded to posts on beams that need to be traversed.

This brings us to the decision on the spacing of anchors for this procedure. Obviously, the anchors S and T must obviously be not farther apart than the distance between the two clamps of the user's double lanyard when stretched full. Alternatively, the combined length of the two lanyards (including the girth of the user) must not be less than the distance between the anchors, paying attention to the need to check if any increased length of lanyards will open the user to any fall risk at any other boundary.

If the clamps are too close and the lanyards are slack during the transfer, if or for other reasons there is the slightest chance that the user can fall from his perch during the transfer, fall restraint may not suffice, and fall arrest measures should be adopted to address the fall risk, as will be discussed later.

4.5   ROPE ACCESS AND WORK POSITIONING

Just as restraint may be used to work in a horizontal plane without fear of falling over an unprotected edge, it may be used also to be applied to prevent falling while working on a steeply sloping or vertical surface. This is called 'rope access' or 'work positioning'.

The difference of this technique from the fall restraint discussed thus far is that now gravity is a major force to address, as the user will be constantly pulled by his weight down the slope or in the vertical drop. The aim will still be to prevent free fall of the user, and give him safe control of his movement, now in vertically instead of horizontally. Further, persons doing rope access and work positioning have to be fully supported externally from the very start.

Both rope access and work positioning involve almost the same considerations, with the difference being that rope access is generally an emergency measure in an unfamiliar situation while work positioning is more a planned and routine activity.

4.5.1   FALL RESTRAINT IN ROCK CLIMBING

To many not familiar with Western culture, it may appear to be a crazy notion for a mature adult to go solo or in a group to climb sheer rock faces just for 'the fun of it', risking life and limb.

During my nearly two decades stay in the USA, I learnt something about not only what drove some of my classmates and colleagues to take up this weekend or vacation hobby, but also how seriously they trained and practiced rock climbing.

The art and science of rock climbing has developed over the decades into a precise organised sport, with so much theoretical analysis and high technology devoted to it, that today, with proper training, it is a much safer sport than many others [Ref. **4.2**].

Rock climbing consists in individuals or small teams climbing steep rock sides of a mountain with the aid of ropes anchored to spikes ('pitons') driven into cracks, and 'cams' that can be opened inside gaps in the rock. Team members are roped to each other, to help and support each other during climbing and in accidental falls.

While in rock climbing, the rope is more used for fall arrest, descending on a rock face in a controlled fashion via a rope, called 'rappelling' or 'abseiling', is a form of fall restraint, equivalent of a person using a rope-grab to adjust his horizontal mobility. Figure 4.4(a) depicts a rock climber rappelling down.

Rappelling applies also to other rope descent activities like rescue from a helicopter, access to a floor of a high-rise building from the roof or a higher floor for rescue, to apprehend a criminal etc.

Thus, climbing down a rock face or a high-rise need not always be a blind risk! Although to most people, being suspended at height by a flimsy looking contraption may seem to be dangerous, the fact is that to those who have no qualms about heights, rope access or work positioning is much simpler, faster, and safer than erecting a scaffold or otherwise reaching the location from the base below.

4.5.2 Rope access

The term 'Rope Access' is often used interchangeably – particularly by the Industrial Rope Access Trade Association (IRATA) – with work positioning.

However, the term 'access' opens up the application of work positioning to moving to and from a desired vertical location.

Rope access is used for purposes other than carrying out assigned work, such as rescuing or otherwise assisting persons suspended or trapped at height and in trouble, hence will involve moving up, down, and sideways.

In order to take care of the dynamic effect and any extra loads (such as the rescuer carrying a disabled or unconscious fall victim), the rescuers will need to be given fall arrest PPE and adequate anchor capacity – both of which will be treated in a subsequent chapter [Ref. **4.3**].

For the rescue, a trained person descends from the roof or other higher level – or a helicopter – to the suspended or stranded person via a cable. He then attaches a separate cable to the victim, or carries him attached to his own cable, to be moved to safety. Figure 4.4(b) shows a rescuer having rappelled down from a helicopter, in the process of being winched up with the victim.

FIGURE 4.4   Rope access and work positioning.

(a) Rappelling down rock face. [Source: Own work. Author: Cans salvamento Galicia. Link: https://commons.wikimedia.org/wiki/File:Rapel.jpg]

(b) Rappel rescue from helicopter. [Source: uploader created. Author: Nic Holmes. Link: https://commons.wikimedia.org/wiki/File:Rapel.jpg]

(c) Work positioning. [Source: Code of Practice for Working Safely at Height, Figure 10.3. Author: With permission from Singapore Workplace Safety and Health Council. Link: www.tal.sg/wshc/resources/publications/codes-of-practice/code-of-practice-for-working-safely-at-heights-english.]

(d) Window washing. [Source: Unsplash. Author: mkjr. Link: https://unsplash.com/photos/2zUjvV0M9dQ]

4.5.3 Work positioning

The fall restraint discussed in the previous section pertained to persons moving around at height in a horizontal plane or over a surface sloping at a small angle less than about 1 vertical to 10 horizontal.

For steeper slopes and vertical drops, the fall restraint is more critical because gravity is already pulling the person down.

In work positioning, access would involve ascending from the base or descending from the roof of the structure to the desired level, carrying out the specified task, and moved back after it is over.

But the principle is still the same: Enable the person to move around and carry out his task without the risk of falling from height from the chosen level. There shall be no fall from the planned position. As with horizontal (or small slope) fall restraint, both his hands are free for him to carry out his task.

This person will be suspended with a chest harness in a comfortable seat belt by means of a working line from an anchor at top, as shown in Figure 4.4(c) [Ref. **4.4**].

He will not fall – it is fall prevention – and the only special requirement is that the user has a head for heights and does not panic when looking down from heights while hanging in space.

Principal requirements are:

(a) A temporary or permanent anchor of capacity about three times the maximum weight of user;
(b) In addition to the working line, a back-up suspension line in case the working line fails.

Of course, other trades also use work positioning to service high-rise structures, such as window washing [see Figure 4.4(d)], facade painting or repair, using battery-operated tools.

4.6 LADDERS

4.6.1 Use of ladders

Ladders cannot be classified under any conventional definition of 'work at height'. They are intended as a means of access and egress for work at heights. If they are used for any task, the task should be light and last not more than about 15 minutes, and special safeguards would be necessary.

Ladders are so common and taken for granted, we do not even notice their presence or who is doing what on them. They can be seen in most offices and shopping centres and (at least in the West) in most homes.

However, people keep falling from them. Most of the falls go unreported because the injuries are not serious enough to go to a clinic or hospital; it is also embarrassing to admit we fell off a ladder!

The number of falls from and accidents with ladders are many times the number of other mishaps at home or in the workplace. As far as employers – and the victim – are concerned, whether one falls from a ladder or a more impressive scaffold, an injury is injury, and the employer's liability is still a liability. The injuries are not always minor either. Lives have been known to have been lost falling from ladders even under 2 m.

Hence, many countries have a number of regulations on how to use a ladder safely and avoid falls.

4.6.2 GENERAL LADDER SAFETY

Fall from ladders cannot effectively use collective fall prevention controls such as guardrails and, hence, require individual fall restraint.

Ladder users fall mainly because they have breached one or more of the safety recommendations given. Often, they lean backwards or sideways so far that their centroid falls beyond the base of the ladder, and if they do not recognise the risk and take precautions, they will fall.

The basic requirement for ladder safety is that the user be trained and disciplined sufficiently to follow all the rules, if needed with the help of a watcher who will monitor the user's activities so that he will not violate any of the rules during his work – this 'buddy' requirement is mandatory in certain codes of practice.

The ladder must be stable, with the bottom and top supported and secured, and the feet level. The universally recognised 'three-point rule' refers to the user always keeping in contact with the ladder stiles or rungs with three out of his two hands and two feet. This means that he should not carry anything in his hand while on the ladder.

Edge protection will obviously not work with ladders, because even if handrails are provided, they will be useful only for holding on while climbing up or down, and when the user slips on a rung, there is nothing to stop him from tumbling downwards!

Actually, many codes classify the sliding anchors as 'fall arrest' and not 'fall prevention' devices on the theory that they become effective only when the user slips a rung or leans too far and falls. But wherever the grab devices are installed, users must necessarily operate them, pressing the clutch and releasing it alternately, and very few of them slip, to be held up by the grab. That is why they have been included here under fall prevention.

Fall arrest solutions may not be feasible for ladders, because – as we will discuss later – fall arrest will involve large impact forces and some means to resist them, all of which will not suit a skeletal stand-alone item like a ladder.

Further, if the user has to carry out any task while standing on the ladder, he must have and use some means of restraining his fall while he uses both hands.

An obvious solution would be fall-restraining lanyards, which can be attached from a frontal D-ring to a rung or other attachment to the ladder. As the user's ladder use zone extends from bottom to top, the fall restraint too must extend the full length of the ladder.

4.6.3 SAFETY FOR VARIOUS TYPES OF LADDERS

(a) Lean-to ladders

Figure 4.5(a) depicts a lean-to ladder with some of the recommended – and in many countries mandated – considerations governing the proper use of such ladders marked on it.

The man at the base of the lean-to ladder, touching it with his fist at the end of his extended arm, is to illustrate a simple trick to check and adjust the slope of the ladder at 1 horizontal to 4 vertical, during setting up.

Usually portable, some lean-to ladders may be fixed to access permanent facilities. Portable ladders are also available in extensible forms allowing them to be used for varying heights.

(b) Fixed ladders

To access work platforms on temporary structures, crane cabins, etc., and to facilitate frequent adjustments and maintenance of equipment or facilities in buildings and on highway gantries, steel ladders are fixed to the structure, generally vertically, as in Figure 4.5(b), but occasionally at a small angle to the vertical.

Invariably above 3 m (≈10 ft), a 'cage' of circular hoops is required attached to the fixed ladder as fall prevention.

The cage hoops will be a great help for side support but obviously will not stop a vertical fall; they are more for the user to hold on to for greater stability and mobility.

Recently, it has been noticed that the hoops themselves may become a source of injury by limb entanglement. If the user slipping on a rung does not stop the fall in time and happens to have an arm or leg slide into a gap between the hoops, the arm may get broken or dislocated, and the groin area between the legs will get badly damaged!

As of November 2016, OSHA declared phasing out cages over the next 20 years in favour alternative methods such as steeply sloping steps with handrails or other fall control safeguards.

(c) Stepladders

Stepladders of the type shown in Figure 4.5(c) are a common item in shopping centres and offices because they are portable and can be erected independent of any support at top. Most of the requirements of lean-to ladders apply to stepladders also.

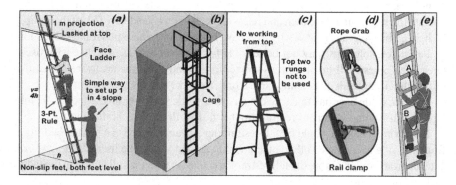

FIGURE 4.5   Ladder use and safety.

Source and Author: Mostly generic items available on the web, rearranged and enhanced with own work.
(a) Lean-to ladder.
(b) Vertical ladder with cage.
(c) Stepladder.
(d) Rope-grab and rail clamp.
(e) Double lanyard as work restraint.

Stepladder users are prohibited from using the top two rungs of the ladder so that they will always have a hand-hold; this definitely implies that the user shall not sit or stand on top of the stepladder.

4.6.4   FALL RESTRAINT FOR LADDERS

Fall restraint for work on ladders, especially the vertical and the tall ones, may be one of many forms. Figure 4.5(d) shows two types of sliding anchors: the 'rope grab' and 'rail clamp', in which a grab mechanism operates either outside a rope or cable or inside a metal groove, the rope or groove rail being attached to the ladder stile or along the rungs. The grab is operated by the climber as he ascends or descends by a hand clutch. The rail clamp may have a lock operated by the climber. Both may have an automatic brake in case of sudden jerk due to a fall.

Both of these devices release the user from the obligation of the three-point rule, enabling him to carry out his task with both hands, during which time he may lean back to keep the connecting lanyard taut, adding to his stability.

Figure 4.6(e) shows the use of double lanyard for climbing a ladder, with the climber alternately clamping A and B every two or three steps. This is the method most commonly used by service persons for wind turbines and electrical towers, with tall, vertical ladders.

Of course, the person who uses it must have the stamina and strength to climb hundreds of ladder rungs – with occasional landings in which he can snatch some rest. In addition, he must have the discipline and the will to strictly follow the following drill as he climbs:

(a) He hooks his (say) right-hand lanyard clamp A to a higher rung;
(b) He climbs a few rungs, with both lanyards attached to rungs;
(c) He unhooks his left-hand clamp B from the lower rung and hooks it to a higher rung;
(d) He repeats the steps (b) and (c) until he reaches the top.

After finishing his task, he reveres the process and reaches the base.

REFERENCES FOR CHAPTER 4

4.1   ———. *How to Use a Miller Titan Rope Grab for Both Vertical and Horizontal Applications*. Retrieved on 19 February 2023 from: www.youtube.com/watch?v=0LANLqEB8Q8

4.2   Seifert, L., Wolf, L.P., and Schweizer, A. (Editors), *The Science of Climbing and Mountaineering*. 13 April 2018, 330 pp, ISBN: 978-1138595231.

4.3   ———. "Release and rescue of a co-worker suspended on a long rope", *Petzl International*. Retrieved on 17 February 2023 from: www.petzl.com/INT/en/Professional/Release-and-rescue-of-a-co-worker-suspended-on-a-long-rope-?ActivityName=On-site-rescue

4.4   ———. *Fall Arrest, Fall Restraint, and Work Positioning: A Quick Guide*. QAB Systems. Retrieved on 17 February 2023 from: www.qabsystems.com/2022/02/04/fall-arrest-fall-restraint-and-work-positioning-a-quick-guide/

5 Mechanics of falls and fall arrest

5.1 WHAT HAPPENS WHEN PEOPLE FALL

While in theory all falls are preventable, falls will involuntarily occur in practice. We have already noted situations where falls are unavoidable and even sought. Then, we have to find ways of stopping (i.e. 'arresting') the fall before the falling person crashes to the ground or other base and gets injured or killed.

Many know about dangers in working at height and how to control them. But – except engineers who have had statics and dynamics (and remember them! – few know, or would even want to know, how falls occur, what exactly happens when one falls, and why the consequences of falls vary so widely.

Understanding the mechanics of falls and their consequences will help us manage the aftermath of a fall better.

5.1.1 RELATIONSHIPS BETWEEN FALL HEIGHT, VELOCITY, AND TIME

When any object falls under the acceleration of gravity, the body acquires a velocity proportional to time and drops through a height proportional to the square of the velocity, as tabulated in Table 5.1 for Metric and Imperial.

In the metric system, to convert mass m in mg to force in newtons, g of 9.81 is often rounded off to 10, so that 100 kg = 100×10 or 1,000 N = 1 kN for quick and ready mental calculations.

5.1.2 EXAMPLE 5.1

What will be the height fallen and velocity for an object in 3 s?

Answer:

From the equations in Table 5.1, for $t = 3$ s,

h = $9.81 \times 3^2/2$ = <u>44.1 m</u>, and also = $32.2 \times 3^2/2$ = <u>144.9 ft</u>
v = 9.81×3 = <u>29.4 m/s</u>, and also = 32.2×3 = <u>96.6 ft/s</u>, AND
v = $(29.4/1000 \text{ m/km})/(3600 \text{ s/h})$ = <u>105.9 km/h</u>, and also
   = $(96.6/5280 \text{ ft/mile})/(3600 \text{ s/h})$ = <u>65.7 mph</u>

Using the h found above,

v = $\sqrt{(2 \times 9.81 \times 44.1)}$ = <u>29.4 m/s</u>; and also = $\sqrt{(2 \times 32.2 \times 144.9)}$ = <u>96.6 ft/s</u>, as before.

DOI: 10.1201/9781032648132-5

TABLE 5.1

Formulas for Falling Bodies

Item	Metric	Unit	Imperial	Unit
Acceleration due to gravity, g	9.81	m/s^2	32.2	ft/s^2
Fall velocity, $v = g.t$	9.81 t	m/s	32.2 t	ft/s
-OR-	35.3 t	km/h	21.9 t	mi/h
Fall velocity, $v = \sqrt{(2g.h)}$	4.43 $\sqrt{h}$	m/s	8.02 $\sqrt{h}$	ft/s
-OR-	15.9 $\sqrt{h}$	km/h	5.47 $\sqrt{h}$	mi/h
Height of fall, $h = g.t^2/2$	4.9 t^2	m	16.1 t^2	ft
Height of fall*, $h = v^2/(2g)$	0.051 v^2	m	0.016 v^2	ft
Time of fall*, $t = v/g$	0.102 v	s	0.031 v	s
Time of fall, $t = \sqrt{(2h/g)}$	0.45 $\sqrt{h}$	s	0.25 $\sqrt{h}$	s

* – v in these formulas must be in m/s or ft/s

The simplified expressions in the second and fourth columns of Table 5.1 will also give results close to these.

5.1.3 EXAMPLE 5.2

How long will it take for an object to fall 60 m? What would the velocities be at the end of the fall?

Answer:

Applying equations from Table 5.1, for 60 m fall, time

$$t = \sqrt{(2 \times 60/9.81)} = \underline{3.49\ s}$$

At 3.49 s, fall velocity

$$v = 9.81 \times 3.49 = \underline{34.2\ m/s},$$
$$v = (34.2/1000\ m/km)/(3600\ s/h) = \underline{123.1\ km/h.}$$

5.1.4 CHARTS FOR FALL HEIGHT AND VELOCITY VERSUS TIME

The expressions in Table 5.1 have been plotted separately for Metric units and for Imperial units, in Figure 5.1. For both charts, the left-hand charts are the expanded versions of the portions of the right-hand charts shown shaded near the origin, to facilitate increased accuracy in estimating values.

The fall charts may be used as alternative to calculations from the formulas in Table 5.1, with quick answers of sufficient accuracy. Given any one of the four quantities in the charts, namely time, height, and velocities per hour and per second, the other three may be read off.

For a given time *t*, the height and velocities may be read off as ordinate values of the respective curves on the vertical axis.

Example 5.1 is shown solved as inset in the Metric left figure of Figure 5.1: Start at A, go up to the three curves, and read off on the vertical axes, the three required height value as C (= 44 m), and the velocity values as B (= 30 m/s) and D (= 106 km/h) – close enough to the calculated values.

If height or one of the velocity values are given, say '*y*', start with the *y*-value on the vertical axis, go horizontally to meet the corresponding height or value curve, and go down to find the time *t* for the given value. Then, the other two required quantities may be read off as ordinate values of the respective curves on the vertical axis.

Example 5.2 is shown solved as inset in the Metric right figure of Figure 5.1: Start at A, go horizontally to meet the height curve up at B, and go down to C to find the corresponding time as 3.5 s. From C, go vertically up to the velocity curves and read off the values for D (= 34 m/s), and E (= 123 km/h) – quite close to the calculated values.

5.1.5 PRACTICAL SIGNIFICANCE OF VELOCITIES

Figure 5.2 displays the heights fallen and velocities reached by a falling person over the first five seconds of fall. The bold print distances and velocities are in metric units, and the smaller print italics numbers are in imperial units.

Two immediate practical conclusions should be obvious from the figure:

(a) Rapidity of fall

In the first second – in the time it takes us to say 'Hundred and one', as that is closer to a second of actual time than saying 'one' – the person has fallen 5 m (16 ft), nearly two floors of a high-rise!

What it simply means is that once a person starts to fall, nothing can be done to stop him. Any fall prevention or arrest measures we wish to provide him should have been planned, all prerequisites and corequisites should have been in place, and any personal protective equipment should have been donned by him before he went to that high location and started the task.

In comparison, most other workplace hazards, such as temporary structure collapse, fire, flood, drowning, and confined space toxicity, give some kind of warning and a few minutes or at least seconds for someone to raise an alarm for everybody to leave, and most to escape from the disaster.

(b) Equivalence of fall velocity to vehicle travel

From Figure 5.2, we see that at the end of the first second, the person reaches a velocity of 35 km/h (22 mph); in the next second, he reaches 70 km/h (44 mph).

But what does this mean to the falling person?

One way to understand this situation is to imagine how it would be if a vehicle travelling at that speed should hit the person. Thus, if he should contact an obstacle at 1 s after the fall, he will be hitting it at 35 km/h, which would be as traumatic as a vehicle travelling at 35 km/h hitting him! And in 2 s, it would be as if a vehicle travelling at 70 km/h hit him. It should now be easy to visualise the impact force.

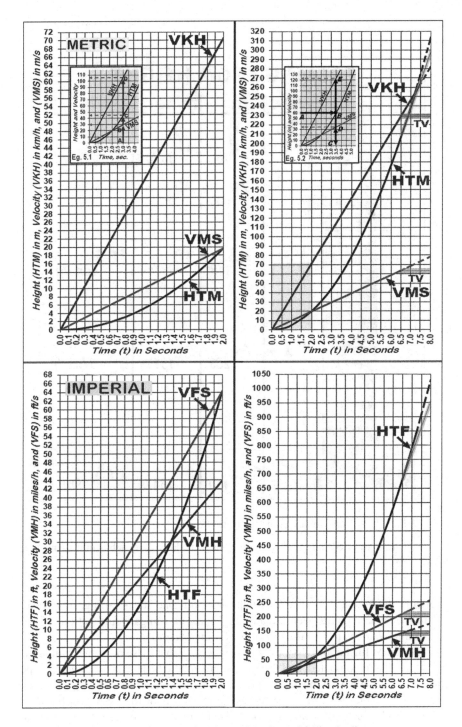

FIGURE 5.1 Charts of height and velocities against time of falling bodies.

Source: Generic knowledge. Author: Own work.
Upper two figures, Metric units charts.
Lower two figures, Imperial units charts.
Left figures of both pairs, enlarged views of right figures.

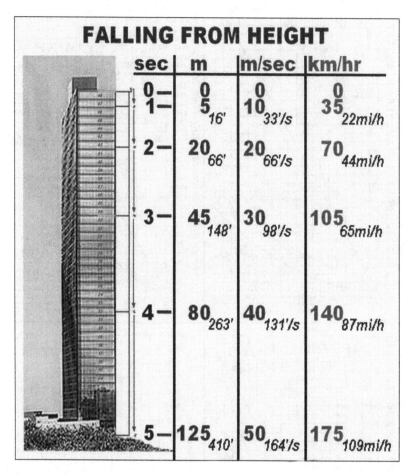

FIGURE 5.2   Height and velocities of falls from 1 to 5 s.

Source: Generic knowledge; high-rise building image modified from public domain picture.
Author: Own work.

Again, the lesson to be learnt is that once someone starts falling, unless well-planned safeguards have been already implemented, things would get out of hand almost immediately, to seal the fate of the falling person.

Fortunately, further discussions here would show that these outcomes can be easily predicted and computed, and fairly standard solutions do exist to protect the person exposed to falls from harm, with reasonable cost and effort.

In many countries, 2 m – commonly taken as 6 ft in the USA – is the height beyond which working at height is required to have fall safeguards. This corresponds to a velocity of 22.5 km/h ($\approx$14 miles/h), achieved in a time of 0.6 s. Even with fall arrest safeguards, free falls are required to be limited to 1.8 or 2 m (about 6 ft).

As mentioned earlier, for smaller fall heights, even at level (zero work height), fall safeguards would be required if the surroundings are hazardous.

5.1.6 TERMINAL VELOCITY

From Figure 5.1, it would appear that fall height and velocity would continue to increase indefinitely with time. Theoretically, that would be the case, but it would not be so in practice.

There is the matter of air resistance of the atmosphere (called 'drag') through which the object falls, slowing down the rate of fall after the object reaches a certain velocity, which is known as the 'terminal velocity', marked in the figures as 'TV'.

The value of this limiting velocity is a function of many factors such as the air density and surface area exposed in the direction of fall.

In general, for falling persons who present a constant profile with arms and legs stretched in a face-down ('prone') position, the terminal velocity is found to be about 65 m/s or 230 km/h (143 mph), which occurs around 6.5 s into the fall, at around 205 m (673 ft).

Naturally the change from increasing velocity to constant terminal velocity is not sharp, but gradual, over a few seconds.

Once the terminal velocity is reached, the object will continue to fall at nearly the same constant velocity, except for changes resulting from wind gusts, changes in the orientation of the body, etc. The height curves will continue as straight lines, as shown dashed in the two charts.

After falling about 200 m (660 ft), whether it is 2,000 m or 20,000 m, the fall velocity will not increase and further falling will not cause more harm – assuming of course that other essentials like oxygen support at heights above about 4,500 m (15,000 ft) are available.

In practical down-to-earth work however, falls may be from zero height to within a few storey heights, less than 100 m.

This topic has been included only for academic interest and is irrelevant to our overall discussion in this book, because our general public or workplace personnel will not be falling from such great heights as to reach terminal velocity.

5.2 IMPACT FORCE AT THE END OF A FALL

It may sound like a joke when we hear: "*The falling is no problem, it is the sudden stopping that hurts!*", but it is the truth!

We shall examine what stopping a fall involves.

5.2.1 STOPPING DISTANCE

Every fall has to end at the ground, lower floor, or other landing area. The fact remains that every moving object must and will travel a certain distance (d) to come to a stop, however small this 'stopping distance' may be.

The stopping distance (say d) cannot be zero, only because zero distance means zero time to stop, and hence, the stopping force on the object will be the velocity just before the stop divided by zero time, which will be infinite, a physical impossibility.

This assertion may also be confirmed by remembering that every material on this earth is elastic to some degree and has a failure limit and will, therefore, elastically deform by a certain amount, and when its resisting capacity is reached will crush, all this adding to the stopping distance.

Linear elastic deformations will be recovered once the load is removed.

This is the reality, the truth.

Note that we are not interested in, and hence not discussing, elastic rebounding due to impact of two bodies, but the collapse of one body against another, coming to rest in a single move.

Let us examine what the stopping distances 'd' are, in a few cases:

- Car travelling at 10 km/h on dry road, good tyres, and brakes, as observed by researchers: $d = 0.4$ m, more if road is wet or slippery, and if tyres and brakes are bad. d varies as the square of the velocity, and hence at 60 km/h, $d = 0.4 \times (60/10)^2 = \underline{14.4 \text{ m}}$.
- Car hitting object: d is the sum of the compression deformations of the two contacting surfaces, which may range from a few millimetres for a low velocity contact, to a metre or more for a high-speed crash. Figure 5.3(a) shows the crushed distance in the crash test of a 2016 Honda Fit, marked 'd'.
- Car hitting pedestrian at high speed, d is elastic deformation of car fender, recovered immediately after impact, plus the crushing of parts of the pedestrian, from a few to many centimetres; pedestrian dies or recovers after some period, depending on parts injured.
- Man catching fly ball: d is elastic deformation of his palms, plus the pullback swing distance of his hands bent at the elbows, plus negligible ball deformation.
- Man rolls off bed: d equals the elastic deformation of the floor and any foot-rug, plus deformation of the body. Usually fully recoverable, unless the elastic limit of the face is exceeded, resulting in a broken nose or worse.
- Man jumps off roof on to driveway: d is elastic deformation of driveway, plus possible crushing of ankle bones. If he is trained to bend his knees and uses it as a shock absorber, flexing them upon contact will provide extra stopping distance; no injury – gymnasts and stuntmen do it all the time! But if he does not bend his knees, he is almost sure to break his ankle and even leg bones.
- Man lands on his head on concrete: d will be the elastic or crushing deformation of the skull plus deformation of the landing surface, which would be a few millimetres at best, usually fatal [see Figure 5.3(b)].
- Man lands on his back on a net: d will again be the elastic deformation of the man's back plus the deflection of the net [see Figure 5.3(c)].

Knowledge about and determination of the stopping distance in each fall case will be critical to assessing the consequences of the fall.

5.2.2 Impact force

When a body of mass m falls a height h and hits the base with a stopping distance d [see Figure 5.3(d)], the average impact force F is, approximately, $(h+d)/d$ times the weight of the body.

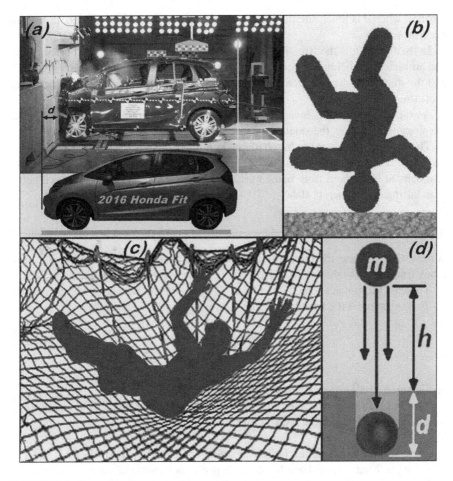

FIGURE 5.3   Stopping distance and impact force.

(a) Stopping distance d in car crash test. [Source: http://www-nrd.nhtsa.dot.gov/database/
VSR/SearchMedia.aspx?database=v&tstno=9337&mediatype=r&r_tstno=9337.   Author:
Calspan Corporation, National Highway Traffic Administration. Link: https://commons.
wikimedia.org/wiki/File:Honda_Fit_-_Impact_Still.jpg. Supervision of car silhouette and
'd' with arrow: Own work.]
(b) Man landing on his head. [Own work.]
(c) Man falling on net. [Superposition of silhouette on generic net: Own work.]
(d) Fall height and stopping distance. [Own work.]

Noting that the potential energy $mg(h+d)$ is equal to the kinetic energy $(mv^2/2)$, we
get, for the average impact force:

$$F = mg(h+\mathsf{d})\,/d = \left(mv^2/2\right)/d \qquad (5.1)$$

from the first part of which, $d = mgh\,/\left(F\text{-}mg\right).$ \qquad (5.1a)

Further, when d is much smaller than h, $(h+d)$ may be taken as h, and Eq. (5.1a) will become $d = mgh/F$.

In the figure, h is the 'free fall" distance, with only the gravitational force acting and air resistance being neglected. Note that the force calculated is the average force, and the maximum force is generally taken as twice this average, but acts for a very short time.

As the impact force is inversely proportional to the stopping distance, the larger the stopping distance, the smaller the impact force.

That is why a cricketer or baseball player pulls his hands back on a catch. That is also why, if a car stops too quickly, the passengers are thrown forward, and it takes seat belts to keep them from being hurt. Likewise, falling on to a net is much better than falling on a concrete slab.

5.2.3 G-FACTOR

A neat way to assess the impact force is with the concept of 'G-Factor', that is, how many times our weight our body feels due to some dynamic event or action.

Thus, in the impact force Equation (5.1), as mg is the weight of the person, the 'G-Factor', namely the multiplying factor on weight, will be given by:

$$G = F/(mg) = (h+d)/d \qquad (5.2)$$

from which
$$d = h/(G\text{-}1). \qquad (5.2a)$$

If d is much smaller than h, then $G = h/d$ and $d = h/G$.

That raises the question, how much G-Force can our body stand, for a short period? There are two surprising facts in this:

(1) A normal adult human body can safely withstand up to 10 G for a short time. Stuntmen and air force/space personnel can take much more with training and some safeguards.

(2) We experience quite a bit of G-force during our daily life and routine activities as follows [Ref. **5.1**]:

Walking – Reference benchmark	... ...	1 G
Average sneeze	... ...	2.9 G
Cough	... ...	3.5 G
A slap on the back	... ...	4.1 G
Rugby, Kabaddi (India)	... ...	5–8 G
Roller coasters, up to	... ...	6.3 G
Plopping into a chair	... ...	10.1 G
Hop, step and jump (Triple jump)	... ...	>20 G

Thus, we may not even notice impact forces up to 5 G and normally will not be seriously or permanently harmed by forces up to 10 G, for short periods.

Limiting the permissible G-factors for humans to 10, it would be more accurate to use $(h+d)/d$ instead of h/d, in our computations.

5.2.4 DURATION OF IMPACT

As the average speed in the stopping distance during the impact will be $(v/2)$, the duration of the impact (t_i), which will generally be of very short time, will be given by the equation:

$$t_i = 2d/v. \qquad (5.3)$$

Obviously, as stopping distance d increases, the stopping time increases, and by Eq. (5.1), the impact force decreases.

Generally, in accidents, the duration of the stopping will be fractions of seconds. This is what we mean by saying that accidents can happen 'in the blink of an eye'!

5.2.5 EXAMPLE 5.3

A person weighing 70 kg falls 4 m on to the ground which (together with his body part compression) gives a stopping distance d of 8 cm. What are: (a) the impact force experienced by him, (b) the impact velocity, (c) the G-factor, and (d) the duration of the impact?

Answer:

We will consider d of 8 cm, as small compared to h of 4 m.

(a) By Eq. (5.1), he will experience a force $F = 70 \times 9.81 \times 4/0.08 = \underline{34.34 \text{ kN}}$.
(b) For a fall of 4 m, the velocity $= \sqrt{(2 \times 9.81 \times 4)}$, or $\underline{8.86 \text{ m/s}}$.
    By Eq. (5.1), the impact force will be: $(70 \times 8.86^2/2)/0.08$, or $\underline{34.34 \text{ kN}}$, as in (a).
(c) G-factor $= 4/0.08 = \underline{50}$, more than he can survive.
    Also, G-factor $= 34.34 \times 1000/(9.81 \times 70) = \underline{50}$, with the 1000 to convert kN to N.
(d) By Eq. (5.3), duration of impact, $t_i = 2 \times 0.08/8.86 = \underline{0.018 \text{ s}}$.

5.2.6 LIMITING G-FACTORS IN PRACTICE

In the opinion of experts on this topic, the maximum G-force that a human body can withstand for a short time without permanent harm is about 5 G [Ref. **5.2**].

As far as possible, workers should not be subject to more than about 2 G on a routine basis or for long durations.

An excellent paper by Sulowski [Ref. **5.3**] summarises the current status on the limiting fall arrest forces in practice. His basic conclusion is that the limiting G-Forces on humans depend on so many variables that no rules are likely to be applicable to all situations.

His recommendations for worldwide adoption are as follows:

- Face-up in a horizontal position, bent at the waist   ... ...   2.75 kN
- Sideways in a horizontal position, bent at the waist   ... ...   4 kN
- Head up, in a vertical position, suspended by a D-ring in harness 6 kN

The most common and worst of the fall positions is the vertical drop, and the 6 kN recommendation have been adopted by many countries.

With 6 kN as our guiding force limit, as male weight varies from 55 to 90 kg globally, the G-Factor would vary from (6/0.55 = 11) G to (6/0.9 = 6.7) G.

However, for routine work at the site, it may be safer to limit the arrest force to a maximum of five times the average weight of the workforce (plus tools) in any region, subject also to a maximum of 6 kN.

We must be extra careful to take these limitations, in providing and ensuring all fall controls, requiring much pro-active planning and rigorous implementation.

5.2.7 How to reduce impact force

From the preceding, it should already be clear that it is not just the height of fall and the consequent velocity that determine the arrest force, but also the stopping distance that is available or is provided.

This gives us a way of controlling the impact force to below harmful levels. All we need to do is to decide the G-factor appropriate to the situation and provide the stopping distance required for the particular fall height.

Naturally, there is a practical limit to the stopping distance we will be able to provide in a workplace. There are of course other ways of diverting or diffusing the high impact forces such as shock absorbers, which will be discussed later.

Alternatively, the free fall may be limited so that it does not exceed the available or feasible stopping distance times the permissible G-Factor.

Thus, if we can get or provide a stopping distance of 0.3 m, and the G-Factor to (say) 4, then, the free fall must be limited to 0.3×4, or 1.2 m.

5.2.8 Example 5.4

What stopping distance will have to be provided for a person weighing 70 kg likely to fall 4 m, to limit the impact force to 5 kN? What will be the G-factor?

Answer:

Considering stopping distance d as of the same order of magnitude as fall height h, Eq. (5.1a) gives:

> $d = 70 \times 9.81 \times 4/(5 \times 1,000 - 70 \times 9.81)$, the 1000 to convert kN to N = 0.637 m (or more).
> G-factor = $(4 + 0.637)/0.637$ = 7.28.
> [Check: $F/(m.g) = 5,000/(70 \times 9.81)$ = 7.28.]

5.2.9 Example 5.5

What should be the stopping distance for the person weighing 70 kg falling 4 m, if he wishes to limit impact to 3 G? What force will he experience?

Answer:

By Eq. (5.2a), stopping distance, $d = 4/(3-1) = 2$ m, which may be provided by means which will be discussed under fall arrest later.

He will experience a force $= 3 \times 70 = 210$ kg, or nearly 2.1 kN, well below the 6 kN limit.

5.2.10 EXAMPLE 5.6

What should be the stopping distance for a person weighing 80 kg (with tools) falling 5 m, if the impact force must be limited to 6 kN? What would be the G-factor?

Answer:

By Eq. (5.1a), $d = 80 \times 9.81 \times 5/(6000 - 80 \times 9.81) = \underline{0.75\ m}$
By Eq. (5.2), G-Factor $= (5 + 0.75)/0.75 = \underline{7.67.}$

5.3 STRESS ON BODY AT IMPACT

5.3.1 STRESS, THE ULTIMATE CONCERN

It is not the height the victim falls from, not the velocity at which he hits the base, not the stopping distance, not the impact force, not even (within limits) the Gs he endures at the impact.

It is whether or not the fall injures him; if injured, whether he lives or dies; if he lives, whether or not he will recover full use of all his limbs and organs, and so on.

The fact of the matter is that injuries happen because the spot where his body hit the base (or other obstacle) broke his skin, drew blood, and/or injured what lay behind the skin at that spot.

(Of course, he could have died of shock, or a massive heart attack, but that we cannot predict or control.)

In terms of engineering mechanics, it is the stress that the person experiences that really harms the body, and not the force. From basic mechanics, we have:

Stress on the body = (Force at point of impact)/(Area of contact)

This stress must not exceed the capacity of the bone, muscle or other part of the anatomy to take it.

5.3.2 STRENGTH OF BONES

Most reported values are only typical averages over all the samples studied, to be taken as general guidance.

Table 5.2, from Ref. **5.4**, gives typical average values for various strengths of adult bones for the outer (Cortical) layer under different loadings, and for the inner (Trabecular) spongy layer under longitudinal load.

Other sources cite skull breaking strength at about 80 MPa. However, the brain is floating around inside the skull in a liquid, and more than exceeding the skull

TABLE 5.2
Strength of adult cortical bone

Nature of Load	Type of Force	Value, MPa
Longitudinal	Compression	193 (50)*
	Tension	133 (8)*
Transverse	Compression	133
	Tension	51
Shear	Shear	68

Note: * The values within brackets are for Trabecular bone.

strength, the damage to the brain due to the impact of the skull could be more traumatic than skull fracture.

To find the load capacity of a bone taking the load, we must know the area of its cortical portion, neglecting the much smaller contribution of its inner spongy trabecular portion.

The bone cross section is often assumed to be an annulus (two concentric circles) with the inner diameter half the outer diameter.

A caution and a disclaimer

Biomechanics is a complex science and involves wide ranges of numerous variables, which depend heavily not only on geographical, ethnic, and genetic factors but also on the individual's anatomy and health history.

All this biomechanics is of great theoretical interest and practical application by experts, but beyond the scope of this book. Use of such data to decide further fall management is not encouraged at this level.

This limited background material is given only to set some related ideas in perspective, and to suggest analytical techniques.

5.3.3 WAYS TO REDUCE STRESS

Reduction of stress will reduce injuries and associated pain, but is subject to G-force limits because of other possible damage to internal organs.

(a) Increase of contact area

It should be clear by now that to reduce the stress due to the impact load, we need to present an increase area to the force. How much this will help may be illustrated by the two falls shown in Figure 5.3(a) and (b).

In Figure 5.3(a), the initial contact diameter area of the head with a hard flat surface may be barely 2 cm, offering an area of about 3 cm^2.

In contrast, falling on one's back as in Figure 5.1(b) presents an area of about 20 cm by 40 cm (at least) or 800 cm^2. Typically, the strands of the net cover from 10% to 20% of the overall area, so taking an average of 15%, area supporting the back will be $(800 \times 15/100)$, or 120 cm², that is, $(120/3)$ or 40 times that of falling on the head, and hence the stress on the body will be down to 1/40th! Note, however, that falling on one's back on a hard surface would cause spinal injury.

Of interest in the above comparison is the fact that falling on one's back on a net not only presents a much large area than other parts of the body, but the net also deflects a lot providing a large stopping distance.

So even if one falls on one's head on a net, in addition to the reduction in impact force due to the net sag, the mesh will wrap around the crown of the head and increase the contact area by many orders of magnitude than on a hard surface.

Many workers in the West (and Australia and other West-oriented countries) train their workers to manoeuvre to drop on their back in case they fall.

This is not as complicated a feat as it appears, because children learn to do it in their P.E. classes at school, and adults who go to fairs and pay to jump from a high platform and fall on to nets or cushions, also learn to do so quite fast.

The rewards are just phenomenal . . . not to forget the saving of one's life!

(b) Presenting a less vulnerable limb

Last but not least, assuming that one can continue to think rationally during the crisis of undergoing a fall from height, it would be better to offer for contact with the base, at the instant of impact, a less vulnerable portion of the body than the head or face, damage to which can be fatal or at least permanently disabling.

Some thought would also go to consider that certain parts of the body are worth 'sacrificing' if only to save more critical parts.

This last principle is quite similar to a fuse in an electrical circuit, which is designed to burn out at a certain amperage so that heavier currents do not damage the entire circuit or sensitive parts of connected equipment.

Thus, falling on one's shoulders, buttocks, or back on to a relatively flat smooth area from low heights like a few metres would bruise the area, but recovery may be total and fast.

Crashing on one's feet would probably fracture the bones of the ankle first and then the bones of the leg, but will spare more vital organs. Likewise, instinctively offering one's hands to prevent smashing the face would break the wrist first, and then the bones of the hand, but save the face.

In both cases, recovery may be complete, although it may take a few weeks. Even if it is partial, it would be better to walk with a limp and lose some writing ability than not be able to walk or write at all!

5.3.4 EXAMPLE 5.6

A worker weighing 90 kg lands on his feet on a concrete slab with an impact force of 25 G. Will it fracture his leg bones if his Tibia's equivalent outer diameter is 23 mm?

Answer:

Although each leg has two bones the tibia and fibula, it is mostly the tibia that takes the load. Tibia's cross section is approximated as annular, with the inner radius half the outer radius.

Outer radius = 23/2 = 11.5 mm; inner radius = 11.5/2 = 5.75 mm.

Area of cross section = $\pi.(11.5^2-5.75^2)$ = 311.6 mm^2.

Force on bone, assuming equal sharing by both legs = $25 \times 90/2$ = 1,125 kg = $1,125 \times 9.81$ = 11,036 N.

Stress in bone = 11,036/311.6=<u>35.4 MPa</u>

This is less than the failure stress of 193(50) MPa in Table 5.1.

Hence, the leg bones are <u>not likely to fail</u>.

(Note, however, the 25 G impact itself, greater than 10 G, would be a major concern for other, internal organs!)

5.3.5 EXAMPLE 5.7

A worker weighing 90 kg falls 5 m and lands on his back on a safety net with a deflection of 1 m. What will be the stress on the back if the contact area is 20 cm by 30 cm?

Answer:

As the stopping distance is of the same magnitude as the fall height, G-factor = (5 + 1)/1 = 6.0

Impact force = 6.0×90 = 540 kg = 540×9.81 = 5,297 N

Overall net impact area = $(20 \times 30) \times 100$ = 60,000 mm^2

Effective net impact area, at 15% = 15% of 60,000 = 9,000 mm^2

Impact stress = 5,297/9,000 = <u>0.59 MPa</u>, about the same order of magnitude as the pressure under our feet while standing, which is around 0.25 MPa.

5.4 SUMMARY OF HOW TO SURVIVE A FALL

The preceding discussion shows that injury depends, in addition to the height (and hence velocity) of fall, on diverse and widely variable factors such as:

- The surface and nature of the base on which the faller lands;
- The part of the body that contacts the stopping surface;
- The PPE and other fall arrest equipment he may be wearing;
- The angle of impact;
- Many other intangibles.

These are among the reasons why it has not been possible to find, and we may not be able to find, a direct or strong correlation between the fall height and the consequent injury.

With all the research that has gone and continues to go into this topic, biomechanics and mathematics can only give broad guidelines and be useful for post-fact explanations than for predictions on fall management.

All that we can do, and should do, is to check over the site of work to find out potential fall locations and obstacles if any in the fall paths and, if we cannot prevent a fall, provide as many relief measures as possible to reduce the fall force and consequent stress on the body, in case of a fall.

The ways to reduce fall injuries may be listed as below, from the following list, in order of preference:

- Minimise free fall height to reduce velocity at impact;
- Increase stopping distance to reduce the impact and/or G-factor within limits;
- Increase area of contact upon contact to reduce stress;
- Present a less vulnerable portion of the body to reduce injury.

5.5 EXERCISES FOR CHAPTER 5

5.5.1 EXERCISE 5.1

What are the heights fallen and velocities reached at 5 s (a) in Metric Units and (b) in Imperial Units?

Answers:

(a) $h = 122.6$ m; $v = 49$ m/s, or, 176.5 km/h.
(b) $h = 402.2$ ft; $v = 160.9$ ft/s, or, 109.7 mph.

5.5.2 EXERCISE 5.2

What are the times elapsed and velocities reached after (a) 50 m fall and (b) 100 ft fall?

Answers:

(a) $t = 3.19$ s; $v = 31.3$ m/s, or, 112.7 km/h.
(b) $t = 2.48$ s; $v = 80.2$ ft/s, or, 54.7 mph.

5.5.3 EXERCISE 5.3

What time does it take and how far will an object fall to reach a velocity of: (a) 30 m/s, (b) 150 km/h, (c) 60 ft/s, and (d) 80 mph?

Answers:

(a) $t = 3.06$ s; $h = 45.9$ m;   (b) $t = 4.25$ s; $h = 88.5$ m;
(c) $t = 1.86$ s; $h = 55.9$ ft;   (d) $t = 3.65$ s; $h = 214.1$ ft.

5.5.4 EXERCISE 5.4

A person weighing 80 kg falls 10 m with a safety harness, which gives a stopping distance of 1.2 m. What are: (a) the impact force experienced by him, (b) the impact velocity, (c) the G-factor, and (d) the duration of the impact?
[Hint: While calculating velocity, remember to convert kN to N!]

Answers:

  (a) $F = 7.32$ kN;   (b) $v = 14.8$ m/s, that is, 53.4 km/h
  (c) G-factor = 9.3;   (d) $t_i = 0.162$ s.

5.5.5 EXERCISE 5.5

What stopping distance will have to be provided for a person weighing 75 kg likely to fall 8 m, to limit the impact force to 4 kN? What will be the G-factor?

Answers:

  $d = 1.80$ m (or more); G-factor = 5.4.

5.5.6 EXERCISE 5.6

What should be the stopping distance for a person who weighs 150 lb falling 20 ft, if he wishes to limit impact to 4 G? What force will he experience?
  [Hint: Weight in pounds is already force units.]

Answers:

  $d = 6.7$ ft; $F = 600$ lb.

5.5.7 EXERCISE 5.7

A worker weighing 70 kg falls on his head three storeys of 3.5 m height on to a concrete floor. He is not wearing a helmet, and the skull contact area may be estimated at 800 mm². If the skull strength is known to be around 80 MPa, and deflection of the skull covered with hair is estimated at 4 mm, will he survive? Why?
  [CAUTION: The values for skull strength, contact area, and deflection are just for illustration, and must not be used in real-life predictions! The outcome will depend on many other factors. – NK]

Answers:

  Stress = <u>2.25 GPa</u>, > 80 MPa. He will not survive. (G-factor of 2,625 is also too high.)

5.5.8 EXERCISE 5.8

If the worker in Exercise 5.7 falls on his head to a net which deflects 1 m on impact, with an estimated overall skull contact area of 7,500 mm², will he survive? What will be the stress?

Answers:

  Stress = <u>7.0 MPa</u> < 80 MPa. He will survive. (G-factor is 11.5, borderline.)

REFERENCES FOR CHAPTER 5

5.1 ———. "What is G-force and what is the maximum value of G-force that you can generate?", *Because Learning*. Retrieved on 23 February 2023 from: https://ehub.ardusat.com/experiments/1597

5.2 ———. "G-Force", *Wikipedia*. Retrieved on 23 February 2023 from: https://en.wikipedia.org/wiki/G-force

5.3 Sulowski, A.C., *How Good Is the 8 kN Maximum Arrest Force Limit in Industrial Fall Arrest Systems?* Sulowski Fall Protection Inc. Retrieved on 24 February 2023 from: www.fallpro.com/fall-protection-info-center/online-articles/maximum-arrest-force-limit-in-fall-arrest-systems/

5.4 Hart, N.H., Nimphius, S., et al., "Mechanical basis of bone strength: Influence of bone material, bone structure and muscle action", *Journal Musculoskeletal and Neuronal Interactions*, 17(3): pp. 114–139, September 2017. Retrieved on 24 February 2023 from: www.ncbi.nlm.nih.gov/pmc/articles/PMC5601257/

6 Collective fall arrest

6.1 SOFT LANDING

Fall arrest is the stopping of a fall before the person hits the ground or other hard surface that will harm him badly and/or permanently. This stopping is basically achieved by the interposition of a stretchable or collapsible medium between the person and the landing surface, to dissipate the fall kinetic energy partly or wholly into another entity or event.

When it is external to the individual and available to all persons falling on it, it is collective fall arrest. The medium thus interposed is termed 'soft landing'.

6.1.1 CHARACTERISTICS OF SOFT LANDING

In case we cannot prevent a fall collectively or individually as described in earlier chapters, we should next aim towards collective fall arrest, namely providing some means of stopping a fall in a planned, less traumatic manner than if the faller hits the ground or other hard base.

The crux of the matter here is that the hard surface will not provide sufficient stopping distance to reduce the impact force to tolerable levels below (say) 10 G, and hence we should interpose some means which will supply the needed stopping distance gradually – this is called 'soft landing'.

Soft landing may be of any material that yields gently and gradually under pressure and (if possible) recovers its original shape when the pressure is removed. Nets, soft mattresses, cushions, shock absorbers in vehicles, etc., are common examples.

The most common soft landings used to stop falls from height are:

1. Safety nets, strung below the anticipated location of the fall landing;
2. Safety cushions made of foam or other filler material, or airbags.

These safe means of arresting a fall should be available to all participants at relevant events and are thus 'collective'.

6.1.2 CONSIDERATIONS IN FALL ARREST

The soft landing materials and designs used should be such that they do not bounce back the objects falling on them, like trampolines or pressurised bags; such bouncing back will be counter to the purpose of terminating falls gently over a stopping distance.

Technically speaking, when an object hits another, there can be various outcomes, depending on the elastic properties and damping characteristics of the objects.

 DOI: 10.1201/9781032648132-6

A tennis racquet is designed to return a ball hitting it with greater or smaller velocity than the incoming hit as the player intends. So is a trampoline designed to bounce back a player higher or lower as the user desires.

Bungee jumpers wish to experience a yo-yo like up and down oscillation a few times, but if it goes on for say, more than half a dozen times, they will get bored and even uncomfortable and irritated. So, the bungee cord must be designed with the appropriate damping to suit.

When a car hits a pothole, the suspension shock absorber is designed to absorb and dissipate the energy so that the car rebound settles down to original stability in very few cycles. Likewise, a trapeze artist falling on the net below wants to quickly swing off its edge and take a bow and not keep embarrassingly bumping up and down.

Considering fall arrest, the intention is to absorb and dissipate the kinetic energy from the high velocity of the fall, in as short a time as possible, and in one gradual braking event. Therefore, devices we use for fall arrest should bring the fallen person to rest in one move.

Then, in most cases, we may estimate the landing impact force on the fall victim or athlete as $(h+d)/d$ times his weight as already discussed.

It is not as if our forebears were ignorant of the need for soft landing. Even Vanuatu elders knew the lads who threw themselves from the tower should land on soft soil. Actually, that principle led to the use of raked earth, loose sand, or saw dust in the landing zones for high jump, long jump, pole vault, and other sports and games where no stakeholder wanted any participant to break a leg!

Figure 6.1(a) shows a pole vault event from the 1912 Summer Olympics at Stockholm, Sweden. While pole-vaulter, American Marc Wright, would have been happy to win the silver medal, the landing area, whatever the 'cushion' was, for his 3.85 m fall in the picture looks hardly inviting!

Nowadays mostly, the earth or sand and saw-dust in sports facilities have been replaced with landing pads of foam 1 to 1.5 m thick, as will be detailed later.

6.2 SAFETY NETS AT PUBLIC FACILITIES

Nets have been in use from early times for fall arrest in various situations in public events.

6.2.1 FIRE RESCUE BY NET

Firemen have been using this technique for centuries to save victims of fires jumping off balconies and windows to escape from the painful death while trapped in the fire. Figure 6.1(b) depicts firemen holding the net invented by Thomas Browder, shown in Figure 6.1(c) [Ref. 6.1].

In the 1970s, as the fire-fighters' ladder technology improved and helicopters could be used to rescue fire victims from higher floors, the Browder net fell out of favour.

However, nets or other soft landing alternatives are still used for fire rescue in many parts of the world.

FIGURE 6.1 Soft landing in public facilities.

(a) Pole vaulting in 1912. [Source: Official Olympic Report, 1912. Author: Photographer of International Olympic Committee. Link: https://commons.wikimedia.org/wiki/File:1912_Marc_Wright.JPG]

(b) Firemen holding a net to catch fire escapee. [Source: Canadian Copyright Collection held by the British Library. Author: William J. Carpenter. Link: https://commons.wikimedia.org/wiki/File:Vancouver_firemen_jumping_into_life_net_(HS85-10-22258).jpg]

(c) Browder's fire safety net. [Source: Own work. Author: Cullen328. Link: https://commons.wikimedia.org/wiki/File:Life_net_Napa.jpg]

(d) Trapeze artistes over net. [Source and author: Circus Acts Australia, with permission.]

(e) Singapore flyer, with safety net at boarding level. [Source: Own work. Author: Dietmar Rabich (1962–). Link: https://commons.wikimedia.org/wiki/File:Singapore_%28SG%29,_Singapore_Flyer_2019_4491.jpg]

6.2.2 CIRCUS NETS

Trapeze acts started in circuses with Jules Leotard doing it for the first time in 1856. He did it over a swimming pool in the beginning, to avoid injury if he fell, and later used mattresses on the ground. In 1866, a safety net strung below the acrobats was used for the first time [Ref. **6.2**].

Today, the safety net is a standard piece of safety equipment during training and performance of almost every aerial circus act all over the world.

A recent article by Renée Fishman [Ref. **6.3**] describes the use of nets for trapeze acts and cautions that nets should not be taken for granted, but must be used in a proper manner, as landing on the net wrongly may bounce the person out of the net, or cause other physical injuries. Figure 6.1(d) shows two circus performers training for the trapeze act, with the safety net below them.

6.2.3 AMUSEMENT PARK NETS

Nets are used as collective fall arrest in public places such as fairgrounds and amusement parks where people may be exposed to accidental falls.

On Sunday, 27 November 2011, at the embarkation zone of the Singapore Flyer Ferris wheel, Figure 6.1(e), a five-year-old boy fell off the boarding platform. Luckily the safety net half a metre below caught him. He was brought back up with no injuries within a few minutes [Ref. **6.4**].

Fortunately fall accidents at public facilities are very rare, compared to construction and other workplaces.

6.3   SAFETY NETS AT THE WORKPLACE

Nets have been used as a safety measure in the construction of buildings and bridges since early 1930s and are now currently standard safety equipment particularly in steel construction.

6.3.1   SAFETY NET UNDER THE WORK

As already mentioned, the first use of safety nets for construction was during the erection of the Golden Gate Bridge in San Francisco, USA, during the 1930s by its engineer Strauss.

The use of safety nets below steel erection is now standard practice as collective fall arrest in most countries with a robust occupational safety and health culture. Figure 6.2(a) shows a net under timber roof construction, and Figure 6.2(b) one for steel construction.

6.3.2   CATCH-NETS AROUND THE PERIPHERY

It is also current common practice to provide catch-nets around the periphery of a work zone at height, primarily to catch falling debris. These may be modified to catch persons if other fall preventive or arrest measures are not feasible for some reason. Figure 6.2(c) displays one such catch-net during construction.

Catch-nets should be as close in height to the unprotected edge as possible, never more than about 1.8 m, extend at least 3 m beyond the unprotected edge, and must not be farther than 225 mm – preferably not more than 100 mm) from the façade [Ref. **6.5**].

These are usually cantilevered out from the structure and need special attention in design and erection.

A net with a mesh fine enough to catch the smallest debris likely to fall may have to be loosely overlaid on the net laid to catch personnel.

6.3.3   DESIGN OF SAFETY NETS

Regional codes of practice and guides provide design criteria for safety nets. Recommended deflections for nets of various sizes and for various fall heights from BS-EN-1263 (Figure 4) are charted in Figure 6.2(d) [Ref. **6.5**]. The chart is applicable to nets (i) whose area is more than 35 m^2, (ii) shorter side is at least 5.0 m, and (iii) initial sag is no more than 10% of the smaller side. An additional 0.5 m should be available below the recommended deflections in the chart.

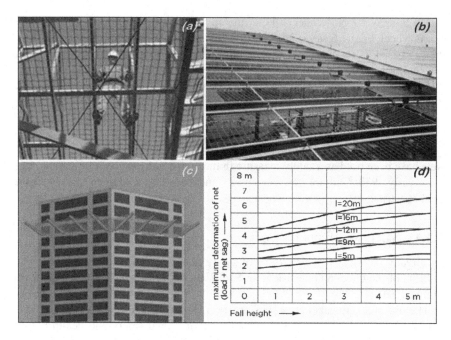

FIGURE 6.2   Safety nets for construction.

(a) Net for timber roof construction. [Source: OSHA Technical Manual (OTM), Section V:
    Chapter 4 – Fall Protection in Construction; Figure 13. Safety net system. Author: OSHA,
    USA. Link: www.osha.gov/otm/section-5-construction-operations/chapter-4]

(b) Net for steel erection. [Health and safety in roof work (Fifth Edition), Figure 33 Use of
    safety nets during industrial roof work. Author: HSE, UK. Link: www.hse.gov.uk/pubns/
    priced/hsg33.pdf]

(c) Peripheral catch-net. [Source: OSHA Technical Manual (OTM), Section V: Chapter 4 –
    Fall Protection in Construction; Figure 12. Safety net system. Author: OSHA, USA. Link:
    www.osha.gov/otm/section-5-construction-operations/chapter-4]

(d) Recommended net deflection. [Source: BS EN 1263-2:2014 – Temporary works equipment.
    Safety nets, Safety requirements for the positioning limits from BS-EN-1263. Author: Brit-
    ish Standards Institution. Link: www.en-standard.eu/bs-en-1263-2-2014-temporary-works-
    equipment-safety-nets-safety-requirements-for-the-positioning-limits/#:~:text=This%20
    European%20Standard%20specifies%20safety,accordance%20with%20EN%20
    1263%E2%80%911%20.]

The following design points are noteworthy:

• The smaller dimension of the net should not be less than 2 m, but may be up
  to 3 m for fall heights up to 6 m;
• The net should be positioned as close to the work level as possible, not more
  than 2–3 m below the work platform level;
• The initial net sag due to the self-weight of the net should be between 5%
  and 10% of the shorter side of the net;
• The side attachments should be 1.5–2.5 m apart;

- The corner and side anchors should be of 6 kN capacity; if there are more than two on a side, they may be alternately 4 and 6 kN;
- The net should preferably be knotless type rather than knotted;
- The mesh should preferably be square rather than rectangular or diamond-shaped, with mesh size 60 to 100 mm.

Safety nets intended for personnel will also catch debris, dropped tools, and equipment. These things generally will pose fresh risks to the persons falling on it. Hence, the safe work procedure should include instructions to stop the work as and when extraneous materials fall on the net and to remove such materials before restarting the work.

6.4 CUSHIONS FOR SOFT LANDING

While nets are indeed an excellent solution to catch falling persons, they do require planning and selection of installation of at least four, or often more, anchor points along their sides to secure them suitably. Their design is complex and their implementation takes time.

Faster, cheaper, and more convenient alternatives to nets for soft landing in many situations are pads or cushions, namely mats filled with foam or other compressible material, or with air, and of such thickness and arrangement that they would sink under load and bring it gradually to rest, this large deformation providing the stopping distance.

6.4.1 MATS OR PADS

We have been laying a cushion around a child's crib or – when needed – even an adult's bed to soften a potential fall. Cases are also known when neighbours bring their bed mattresses and pile them up at the base of a building, to facilitate upper floor tenants jumping on them to escape from a fire.

The same idea is formalised to reduce fall injuries at gymnasia and sports training centres and competitions. Pads are particularly critical where the participants land head first or on their back, as in the now popular backward flip high jump.

Figure 6.3 displays examples of such pads. In the figure, (a) shows a just completed school gymnasium in Germany; (b) is a foam pad for bouldering; (c) is a training centre for gymnastics; (d) shows the high jump landing pad in the 2000 Summer Olympics at Sydney; and (e) depicts the pole vault set-up at the 2006 Commonwealth Games at Melbourne, Australia. Their thicknesses reflect the best estimates of the landing forces to be expected.

However, the problem with pads is that while they are useful in fixed locations, they are bulky, limited to about 1 m thickness and cannot be moved around easily and quickly from place to place. Hence, they are not practical for workplace falls.

The difficulty of handling huge foam cushions has been solved by proprietary cylindrical foam-filled bags, about 700 mm diameter and 2.5 m longer, attachable to one another to form any desired size of soft landing, even in multiple layers.

FIGURE 6.3   Foam pads for sport and training.

(a) School gymnasium in Germany. [Source: Own work. Author: Bauken77. Link: https://com-mons.wikimedia.org/wiki/File:2018_neue_Ger%C3%A4tturnhalle_der_Mittelschule_Wolfurt_und_TS_Wolfurt.jpg];

(b) Foam pad for bouldering. [Source: Own work assumed (based on copyright claims). Author: Romary assumed (based on copyright claims). Link: https://commons.wikimedia.org/wiki/File:Crash_pad.jpg]

(c) Training centre for gymnastics. [Source: Own work. Author: Martin Rulsch. Link: https://commons.wikimedia.org/wiki/File:2019-06-26_1st_FIG_Artistic_Gymnastics_JWCH_Men%27s_Training_26_June_Afternoon_2_Still_rings_%28Martin_Rulsch%29_084.jpg\

(d) High jump landing pad in 2000 Summer Olympics at Sydney. [Source: www.flickr.com/photos/thepaperboy/1488165716/. Author: Ian @ ThePaperboy.com. Link: https://en.wikipedia.org/wiki/File:Clearing_the_bar_in_the_men%27s_high_jump_final_Sydney_2000.jpg]

(e) Pole vault set-up at 2006 Commonwealth Games at Melbourne. [Source: www.flickr.com/pho-tos/mrpbps/2868896039/in/photostream/. Author: www.flickr.com/photos/mrpbps/. Link: https://commons.wikimedia.org/wiki/File:Pole_Vault_MCG,_2006_Commonwealth_Games.jpg]

6.4.2 Airbags

The elegant alternative to a foam cushion or spring mattress is the airbag – also called an 'airpad' – which eliminates all the disadvantages of a solid cushion. This is a sac, usually rectangular in plan and when filled, of constant thickness, manufactured in standard sizes from $6 \times 8 \times 4$ ft thick to $10 \times 14 \times 6$ ft thick, but almost any other size, on order. Airbags are available for fall height ratings of 20–200 ft (6–60 m).

In fact, airbags are employed in giant sizes covering many hundreds of square metres, for snow-ski training and simulated sports fun.

Individual airbags may be attached to one another on all their edges with click-able buckles to cover any given area without gaps. A group of modular airbags is preferable to one large-size airbag, because the former gives versatility of plan shape and convenience of setting up and moving around.

The unique feature of these airbags is that they will not be air tight and filled bursting tight to capacity, because in such a case, the taut surface of the bag will bounce the striking object back like a trampoline would. That would defeat the purpose of the fall arrest and instead aggravate the situation with the unpredictable rebound causing havoc to the victim.

The airbag will have vents along its sides, with specially designed baffles acting like valves to bleed out air when an object drops on it, gradually slowing down the fallen object to a stop, like a shock absorber. The expelled air will be replenished subsequently by pumps to continue to keep the bag inflated.

Airbags are not unduly expensive. They are portable, reusable, easy to set up and maintain, and except for the very large area bags, they are quite easy to transport and move around.

Another very effective application of the airbag is its use to lift objects as heavy as ships – which of course are outside the scope of this book.

6.4.3   Airbags in public facilities

(a)   Airbags for fun

In most Western and other advanced countries, airbags are so common that they are a popular attraction at amusement parks, patrons including children paying hefty sums for jumping from heights and experiencing the thrill of the jump and the landing, with photos or videos recording their feat, without any unreasonable or unpredictable risk. Many families rent airbags for children-related events or buy their own airbags for private family fun.

Figure 6.4(a) shows one such airbag, with the basic components marked. Obviously, at an amusement park, the paying visitor is taken to a desired height H by a scissor hoist or other means, given some basic instructions, and allowed (sometimes nudged!) to jump. An intermediate tumble position and the final supine fall on the back are shown in the figure.

The airbag consists of one or more layers of airtight compartments with an inlet for air pump to initially fill the sacs and then maintain a certain working pressure. Baffle valve B allows the air to bleed out at a controlled rate when anything falls on the airbag and the object starts sinking into the stopping distance.

(b)   Airbags for training

With the design and controls described earlier, the potential for the use of airbags for training and practice for safely jumping from height for competitions or other reasons would become clear. In recent decades, airbags have replaced the sand or sawdust pits for soft landing at training centres for everything from mountaineering to calisthenics and Olympic gymnastics.

Foam pads are an excellent means to eliminate or minimise harm. They are used extensively and exclusively in most sports and gymnastics training establishments and competitions.

One might attribute most of the reduction in training accidents and at least part of the improved achievement records, to such advances in training facilities.

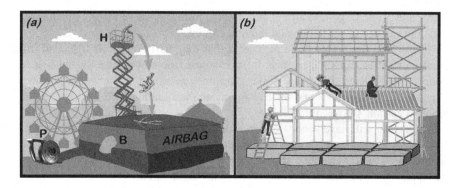

FIGURE 6.4   [Source and Author: Own work, with a few images from public domain sites.] Airbags for soft landing.

(a) Airbag for fun. [Amusement Park background image by Jemastock on Freepik]
(b) Airbag in construction. [Timber framing image from Freepik.]

In fact, huge airbags are spread out over wide expanses in ski-jumping and other extreme sports training facilities, at which eager amateurs try their dreams and serious professionals hone their skills.

Apparently, there are very few restrictions on the size of the airbags. As they become bigger, they become more sophisticated in their design and use.

(c) Airbags for rescue

The portability and ease of setting up the airbags make them ideal for rescuing persons trapped in a burning building, threatened by criminals in higher floors, or otherwise forced to drop from heights.

Organisers of 'Incheon' 17th Asian Games in 2014 trained firemen in rescue from athlete's quarters with airbags placed under windows for the victims to jump on to.

Singapore's Civil Defence Force (SCDF) uses it routinely to catch suicides threatening to jump out of balconies, or criminals planning to escape from windows, of high-rises. A video [Ref. **6.6**] of its rescue of a suicide victim who jumped from 17 storeys above the ground by spreading a large airbag under the ledge on which the young man perched and jumped from is dramatic proof of its efficacy.

6.4.4   AIRBAGS AT THE WORKPLACE

With all the advantages of airbags for saving lives and major injuries from falls, airbags should be the fall arrest soft landing of choice, before considering nets or other fall arrest devices, at workplaces.

Many countries are already using them to great effect, improving productivity with economy. Many construction companies invest in buying a number of manageable units approximately 2 m long and 1 m square in cross section when inflated, each costing a few hundred dollars. They are eminently reusable many times with maintenance and care during use.

Figure 6.4(b) shows a suggested used of compact airbag modules of about 1 m by 1 m cross section and 2 m or more in length, clipped together to offer a large fall zone below the work area with no edge protection or work restraint to the workers, who have only helmets and visible jackets for PPE.

The figure shows one unit under the ladder, and a group of jointed modules along the front of the construction, to catch the two workers shown, if they should fall. If any of them falls on the bags, he will just get up and get back to work with nary a scratch! He should of course learn the lesson that he should not get into a similar situation again because there may be no airbag to cushion his fall next time!

Airbags are best where either there is not enough fall clearance for the use of safety harness or when anchor fixtures are infeasible or not available for other fall arrest PPE. They are so versatile in application that they may even be laid out around a truck bed to save worker injuries during loading and unloading, and under gaps in scaffold or formwork during erection or dismantling.

Airbags are also eminently suited for 'short' falls between 2 m and about 5 m, especially for falls through gaps in floor formwork, falls through fragile ceilings and portico coverings, as explained in my paper [Ref. **6.7**].

6.5   CARDBOARD BOXES FOR SOFT LANDING

6.5.1   BASIC PRINCIPLES FOR FALL ARREST

Surprising as it may sound, cardboard boxes stacked in a pile work quite well to cushion falls from heights. The common cardboard box in which we get most of our products and appliances is a remarkable life saver in falls. I include this option seriously at the risk of being considered foolhardy, hoping it may serve as a fall-back (pun intended!) or ad hoc damage control.

In fact, skydivers, high jumpers, and stuntmen use cardboard boxes more than airbags or nets. Stuntmen in India and many other countries use them exclusively, because cardboard boxes are cheap, easy to get and assemble, and the uncrushed ones can be reused. These will be reviewed in a later chapter.

In high jumping and stunts there will generally be some horizontal motion also, and hence stopping distance must be allowed for in the horizontal direction also, in addition to the vertical drop, to dissipate the kinetic energy safely.

Rap/pop musicians today also need them to absorb their fall when they jump and teeter around on the high stage in their ecstatic frenzy – the cardboard box layers around the stage will be covered with fancy drapes, that is all.

Airbags are designed only to catch humans. Catching persons riding motorcycles, or wearing sharp PPE, would damage the airbag. But cardboard boxes that get damaged by any object can just be thrown away. Actually, it is their crushing that saves human lives – remembering that anything that helps dissipate the kinetic energy ($mv^2/2$) of the fall (or otherwise moving) velocity (v) would mean that much less impact force on the person!

Needless to say, the impact will not be 'soft' like on a net or airbag, but quite hard in fact. The jumper is expected to fall on his back, protected by thick clothing. If he leads with his feet, his boots will take the brunt of the impact. If he is likely to fall headlong, he should be helmeted. In any case, the crushed first box will become the shield for further downward or forward movement, protecting the rest of the body from bruising.

After all, a bleeding nose or a scratched face would any day be better than a major injury or death!

6.5.2 VALIDATION OF CARD BOX USE

Lest readers think that use of a cardboard box is a primitive or unscientific home-grown solution to a serious problem, it is only fair to note that considerable research has been invested in the subject and there are numerous publications on it.

A landmark research report [Ref. **6.8**] presents an equation for the energy needed to crush cardboard boxes of given dimensions and Edge Crushing Strength (ECT) value. From this, the number of boxes needed to arrest a person's fall may be found.

As the exact location of the landing contact point cannot be determined, a certain additional area around the predicted landing point – say 1 m – has also to be filled with the same protective boxes. Further, an additional layer of boxes beyond the theoretically needed amount may also be provided as safety margin.

For an object of mass m travelling at velocity v, the kinetic energy is $mv^2/2$. This energy will have to be absorbed by the crushing of cardboard boxes. From this, the number of boxes required to dissipate the impact and bring the person to a stop can be computed.

There have been many documented demonstrations of persons jumping from planes and landing safely on cardboard boxes. Many of these demonstrations have been skydivers who jumped from helicopters or planes and hence had a horizontal velocity also in addition to the vertical velocity, when they landed. Some of them launched themselves into space from a ledge at height so that they too had a horizontal velocity.

Hence, their stopping trajectory not only was vertical but also had a horizontal component, their length depending on the magnitude of their horizontal velocity v_h.

The determination of the horizontal stopping distance followed the same energy transfer principles as for the vertical fall arrest. The horizontal energy component E_h would be $m.v_h^2/2$, and this also has to be dissipated by the crushing of the required number of boxes horizontally. All this of course, is beyond the scope of this book!

6.5.3 EXAMPLE 6.1

If the crushing energy for one 30 in. ×30 in. ×30 in. (0.76 m × 0.76 m × 0.76 m) standard corrugated cardboard is known [Ref. **6.8**] to be 633 joules, how many boxes will a worker weighing 70 kg falling 3 m need for a safe stop?

Answer:

Energy developed by falling worker

$$= mgh = (70 \times 9.81 \times 3) = 2{,}060 \text{ joules. (1 joule} = 1 \text{ N travelling 1 m)}$$

No. of boxes needed vertically = 2,060/633) = 3.3, that is, 4. (We may add one more layer for safety margin.)

He would be expected to land flat on his back, so that we will need a length of (say) (6 ft/30 in), or 3 boxes.

With a safety margin of about 1 m all around, extra (2 m/0.76) or say 3 boxes in each direction, a stack (3 + 3) or 6 boxes long, by (1 + 3) or 4 boxes wide, and (4+1) or 5 boxes deep should be provided.

Total number of boxes = 6×4 ×5 = 120, most of which will not be crushed and can be reused.

6.6 EXERCISE FOR CHAPTER 6

How many cardboard boxes will it take to plan to stop a person weighing 80 kg falling with a velocity of 9 m/s if the crushing energy of one cubical box of 25 cm side is 400 joules?

Answer:

For him to fall on his back and be surrounded on the sides for 1 m, he will need a stack 16 boxes long, 10 boxes wide, and 10 boxes deep, that is, 1,600 boxes.

REFERENCES FOR CHAPTER 6

6.1 Schlags, M., "Fire department safety nets . . . Did they go away and why?", *My Firefighter Nation*, 2011. Retrieved on 25 February 2023 from: https://my.firefighternation.com/forum/topics/fire-department-safety-nets-did-they-go-away-and-why#gref

6.2 ———. "Trapeze origins", *Vertical Wise*. Retrieved on 25 February 2023 from: www.verticalwise.com/trapeze-origins/

6.3 Fishman, R., "The safety net is an illusion", *My Meadow Report*, February 2018. Retrieved on 25 February 2023 from: https://mymeadowreport.com/reneefishman/2018/safety-net-illusion/

6.4 Tan, J., "5-year-old in stroller falls off S'pore flyer platform", *Yahoo!news*, 2 December 2011. Retrieved on 6 July 2023 from: https://sg.news.yahoo.com/5-year-old-in-stroller-falls-off-s%E2%80%99pore-flyer-platform.html

6.5 ———. "Safe use of safety nets", *WorkSafe Best Practice Guidelines*, New Zealand Government, May 2014, 31 pp.

6.6 Ang, M., "Teenager fell off 17th storey ledge in Sengkang, saved by SCDF air pack", *Mothership News*, 1 March 2019. Retrieved on 27 February 2023 from: https://mothership.sg/2019/03/sengkang-17th-storey-attempted-suicide/

6.7 Krishnamurthy, N., "Simple way to save lives in low falls", *Safety Matters*, Singapore Institution of Safety Officers, January–March 2020, pp. 13–179. Retrievable from: www.profkrishna.com/ProfK-Publications/Publications.htm

6.8 Jeffrey, G., Esser, E., and Pai, S., *Cardboard Comfortable When It Comes to Crashing*. Retrieved on 17 August 2017 from: https://sites.math.washington.edu/~morrow/mcm/uw24.pdf

7 Individual fall arrest

7.1 AIM OF INDIVIDUAL FALL ARREST

7.1.1 NEED FOR INDIVIDUAL FALL ARREST

If and when collective prevention, individual prevention, and collective fall arrest are infeasible for some reason(s), only then shall individual fall arrest measures be employed. It would be the last in the hierarchy of fall management procedures.

7.1.2 ROPE AND WAIST BELT

The conventional approach to fall arrest until the 1970s was a rope tied around the waist or attached to the waist belt, as in Figure 7.1(a).

The rope's length would naturally be shorter than the height of the worker's feet above the base so that no part of him would hit the ground.

Theoretically, it made a lot of sense and worked satisfactorily for relatively small falls – say one or two storey heights. But, for larger heights, the jerk at the end of the fall crushed the victim's internal organs and damaged or broke his spine, after which, he ended up paralysed or dead.

Note that the entire impact force must be resisted at the connection point V by the backbone (shown superposed on the mannequin), which is quite weak in bending.

Moreover, if worn loose, the loop around the waist tended to slip over the rest of the body and deny the wearer further protection from the fall.

By the 1970s, the solution to soften the impact was found in what is today known as the 'full-body safety harness with shock absorber', Figure 7.1(b,c), described in the next section.

With the positive experience gained from the safety harness, America banned the waist belt for fall arrest in 1998. Other countries followed suit, soon after.

But the rope and waist belt – augmented with a chest harness if found necessary – are still staple safeguards for fall restraints and work positioning, because in these instances there will be no fall and hence no jerk impact.

7.1.3 FULL-BODY SAFETY HARNESS

The safety harness had two major innovations that overcame the deficiencies of the rope and waist belt, as depicted in Figs. 7.1(d,e):

(a) It supports the body weight at the heavy-boned shoulders (P) and hips (Q), rather than at the fragile spine (V) in Figure 7.1(a);

(b) It has a shock absorber, marked S in Figs. 7.1 (d,e), which will stretch like a spring from about 15 cm to about 1.2 m when jerked with an impact force of 4 to 6 kN, acting as a gradual brake so that the falling body will not be subjected to sudden harmful impact.

DOI: 10.1201/9781032648132-7

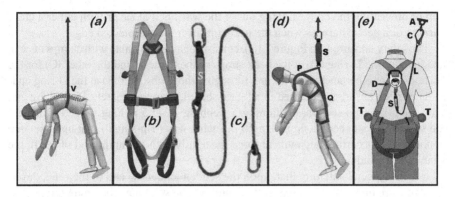

FIGURE 7.1   Waist belt and full-body harness.

(a) Rope and waist belt. [Source and Author: Own work.]

(b) Full-body safety harness. [Source and author: Generic item.]

(c) Shock absorber. (Source and Author: Generic item.)

(d) Safety harness. [Source and Author: Own work.]

(e) Back view of man wearing safety harness. [Source and Author: Own work, based on image from WSHC, Singapore, with permission.]

7.2   PARTS OF FULL-BODY SAFETY HARNESS

7.2.1   THE SAFETY HARNESS

The harness itself, Figure 7.1(e), basically consists of five or six straps of webbing, two to go around or over the shoulders, two around the thighs, and one more across the torso or two across the chest and waist. Readers may refer to my paper [Ref. **7.1**] for details.

As already mentioned in the previous section, the first safety feature of this harness is that the shoulder straps and thigh straps transfer the victim's weight to the anchor, from around the largest bone masses in the body, namely the shoulders and the hips, in contrast to the older way of the waist belt, which forced the back bone to take the entire load and impact, leading to injuries.

Connection of the harness to the lanyard is made through the D-ring, named so because of its shape, usually fitted at the back (dorsal area) between the shoulders, but occasionally, at front as in rope access, or when the wearer has to carry out a task on a vertical or steeply inclined surface.

7.2.2   THE SAFETY LANYARD AND SHOCK ABSORBER

Currently, the most popular individual fall arrest device is the full-body safety harness. While many assume (in the fall arrest context) that the safety harness comes with a 'safety lanyard' which incorporates a shock absorber, the West prefers to use the two terms distinctly. Figures 7.1(b,c) display the two devices.

The reason for this separate usage of terms is that the body harness without the shock absorber may be used for fall restraint, where no fall is involved, with the sole aim of preventing the body slipping out of the waist belt alone or with either a chest harness or a pelvic harness when the user slips over an unprotected edge.

The safety lanyard, L in Figure 7.1(e), consists of a rope or cable with clamps at each end, one for the D-ring (D) at the back straps of the harness, and the other (C) for the anchor (A), or other anchoring fixture, through a shock absorber (S) at the D-ring end.

The shock absorber, enclosed in a fabric sheath, is approximately 15 cm long and 3 by 4 cm in cross section, containing a webbing 1.2–1.5 m long, folded to fit into the sheath and stitched along its length in such a way ('rip-stitch') that upon sudden application of a certain deployment force (generally 6 kN) as at the end of a fall, the stitches tear gradually.

It is this special stitching that, upon the sudden jerk at the end of the rope, develops the stopping distance equal to the length of the webbing, absorbing the impact energy and thus reducing the G-force on the victim to tolerable levels. This reduction of impact force is the second safety feature of the safety harness with shock absorber.

Once deployed, the shock absorber should not be reused, but discarded or sent back to the supplier for replacement.

When there is need for 100% tie-off, a double lanyard may be attached to the shock absorber instead of a single lanyard.

The use of a full-body harness for fall arrest carries with it the inconvenience and discomfort of chafing straps, particularly around the thighs, although the thigh straps are essential to distribute the impact force to the hip bone.

7.3   SUSPENSION TRAUMA

Safety harness carries with it a peculiar risk, arising from the simple act of being suspended from it after a fall.

It may be believed that when a person wearing a safety harness falls, he will be safe and injury-free in the hanging position until a rescue team arrives and manages to get him down or up from wherever he is hanging, to a safe zone.

Unfortunately, human biology reacts differently. Statistics affirm that one in about six persons suspended from a safety harness and continuing to hang for more than an average of from 5 to 20 minutes is open to risk of heart failure.

Actually, even when a person is standing still in the same position for long time, it is likely that he may suffer reduced blood supply to the heart and faint, or even die, in about 16% of the cases, according to medical findings.

This problem is called 'Orthostatic Intolerance', and 'Venous Pooling', and specifically in the case of fall arrest, 'Suspension Trauma' [Ref. **7.2**].

The reason is that, although blood from our heart can reach our feet by gravity, it cannot be sucked back upwards by the heart – it is not powerful enough – for the necessary oxygenating purposes. Special muscles in our legs pump the blood back to the heart, and they stop working when the legs are stationary for a long time.

That is why if a fallen person is allowed to stay suspended in a safety harness for too long, he may faint, and if not rescued but continues in the vertical position, he may die. (In the case of the guardsman, once he falls, the blood circulation from the legs will automatically start again and he revives soon.)

7.4   RESCUE FROM SUSPENDED HARNESS

7.4.1   RESCUE MEASURES

It is clear that while the deadly impact of the falling person when he crashes to the ground or other base has been averted by the use of the safety harness, he cannot be allowed to stay suspended for too long. He must be extracted from his suspended position within about 20 minutes to avert the 16% likelihood of death.

This problem must be addressed as follows:

(1) Call the Fire Department or other emergency service to rescue the victim;
(2) Rather than depend on public services which often may be over-stretched and may not arrive soon enough, pre-plan in-house rescue, by:
  (a) Using proprietary rescue equipment, such as telescoping rods with hooks at the end, to snare the fallen person's D-ring and pull him up;
  (b) Using a scissor hoist, boom hoist, or other MEWP available on site, or can be borrowed from a neighbouring project;
  (c) Using basic rescue devices such as portable plastic ladders (which can be rolled up and carried in a duffel bag) capable of taking the weight of two persons, for the contingency that a rescuer may have to go down and carry an inactive victim to safety or to hook him up to a second lifeline for moving;
  (d) Providing a soft landing (net or airbag) at a stable location below the suspended person for the person to release himself from his D-ring attachment and drop, assisted if necessary, by a rescuer.
(3) With rescue pending, if the person is conscious and responsive, attempt palliative measures to maintain or revive blood circulation, by:
  (a) Having the victim use pull out the foot stirrups (– if previously obtained, as it is not mandatory) in pouches – such as T in Figure 7.1(e) at the two sides of the safety harness belt – on the waist strap of his safety harness – and press his feet on them, so that the leg muscles will begin to start pumping again;
  (b) Dropping a well-anchored rope with a loop at the end to slightly above the victim's foot level, and asking him to place both his feet into the loop and press down;
  (c) Asking him to keep his feet moving, or push gently on any nearby object.

7.4.2   PRECAUTIONS AFTER RESCUE

The problem is not completely over even after rescue, particularly if no steps have been taken to revive the slowing down of blood circulation.

As soon as the victim is extracted from his hung position and brought to a safe place, the victim should not be placed horizontal as would be the normal procedure. For then, the accumulated blood from the feet would rush to his heart and give him a massive heart attack.

He may be made to sit, leaning against some support, and then over a period of some 15 minutes, slowly lowered to a horizontal position in stages.

7.5 COREQUISITES FOR HARNESS USE

For its safe and correct use, the full-body safety harness requires a number of coreq-uisites, which also would be the reasons we should not simply issue safety harnesses for those who do not actually need them.

7.5.1 CORRECT FIT

Somewhat surprising to the uninitiated, and often ignored by those who do not understand the significance of the act, is the imperative need for the safety harness to be worn properly by the user.

Of course, the straps must not be so tight that they would affect the blood supply to the limbs. But they should not be too loose, as is often the case with the ignorant or negligent.

What is required is for the wearer to adjust the five or six straps of the har-ness around his shoulders, thighs, and torso, to a 'snug fit', usually checked by the 'two-finger test', referring to the ability to slide his two (some would go up to four, in flat position) fingers between every strap and the body.

A recent fad is to use the 'pinch test', by tightening the strap gradually till the wearer cannot pick a portion of the strap between two fingers. The problem with this may be over-tightening, because even then one may be able to forcibly pinch them!

The reason for this precaution is that if the straps are loose, the jerk at the end of the fall will force the strap pairs at the shoulders and the groin to come together with considerable lateral force.

While the shoulder blades will be able to resist this lateral force with no problem, the snapping of the thigh straps will, for males, crush their sensitive private parts. Too many male victims have contended with such injuries to ignore the problem.

Hence, wearer training should focus on correct adjustment of all the straps every time the harness is used.

7.5.2 FALL CLEARANCE

A very important criterion for the use of safety harness is that the clear distance below the safety harness anchor must be sufficient for the jerk on the lanyard at the end of the fall to deploy the shock absorber and develop the stopping distance, with-out at the same time allowing any part of the wearer hit the base. Figure 7.2 shows the stable and fallen positions of a worker at height.

The various contributions to the total fall height at the end of a fall are listed in the figure, from which we may write the expression for the minimum fall clearance (C_A) below the anchor, required for a person to effectively use a safety harness, as:

$$F_c = L + a + s + h + m. \tag{7.1}$$

With typical values $L = 1.5$ m, $a = 1.2$ m, $s = 0.1$ m, $h = 1.7$ m, and $m = 1$ m, we get:

$$F_c = 1.5 + 1.2 + 0.1 + 1.7 + 1 = \underline{5.5 \text{ m}}, \text{ a little over } \underline{18 \text{ ft}}.$$

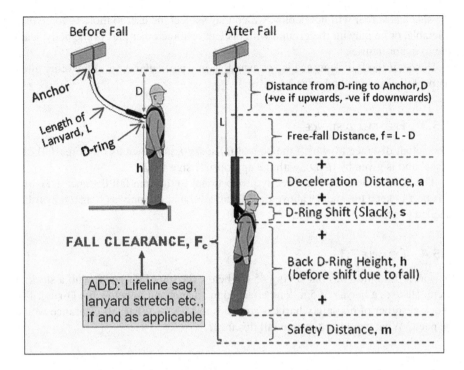

FIGURE 7.2   Fall clearance for safety harness.

Source and author: Own work, modified from Figure 18 of public domain OSHA document, "Technical Manual (OTM) Section V: Chapter 4 – Fall Protection in Construction" by OSHA.]

These values are often cited as the minimum clearance required for use of a safety harness. The actual clearance will mainly depend on how much higher or lower the D-ring is from the anchor and the length of the lanyard.

   In addition to these items, the following must be added to obtain the final value of fall clearance:

   (i)   Sag of any lifeline to which the lanyard may be attached;
   (ii) Stretch of lanyard and of any vertical lifeline if attached to it;
   (iii) Elastic deformation of any post or structure to which the anchor is attached.

Unless significant in magnitudes, the preceding items may be assumed to be included in the safety margin, m.

   From Eq. (7.1), we may get the fall clearance below platform, F_p as $(F_C - D - h)$, noting that D is positive when anchor is above the D-ring, and negative when below, because when the anchor is below the D-ring, the user will fall more than when the anchor is above the D-ring.

   Fall clearance is quite critical and needs to be checked before the safety harness and lanyard are issued for use, because, if the required fall clearance is not available, the shock absorber will not deploy, which may subject the user to more G-force than tolerable, or he may hit the ground with his feet or some other part of the body, leading to major injuries.

Hence, if the fall clearance is not available in a particular case, some other alternative fall arrest method must be utilised.

7.5.3 FREE-FALL DISTANCE

The actual distance by which the person falls, say f, is known as the 'Free-fall distance' and is given by $(L-D)$, with the appropriate sign for D.

The impact force will be directly proportional to the free fall distance f. Hence, it is best to keep the anchor above the D-ring level and use only as long a lanyard as needed to accomplish the task.

7.5.4 EXAMPLE 7.1

A worker 1.8 m tall has his D-ring 45 cm below the top of his head, with a slack of 5 cm. He uses a lanyard 3.5 m long attached to an anchor 1 m below the D-ring. The shock absorber of his safety harness stretches to 1.25 m. What fall clearance would he need? What would be the free fall distance?

Answer:

With the notation of Figure 7.2, we have: $L = 3.5$ m, D (with anchor below D-ring) $= -1.0$ m, $a = 1.25$ m, $s = 0.05$ m, $h = 1.8-0.45 = 1.35$ m, and $m = 1$ m.

Hence, by Eq. (7.1), minimum fall clearance below the anchor:

$$F_c = 3.5 + 1.25 + 0.05 + 1.35 + 1 = \underline{7.15 \text{ m.}}$$

Then, minimum fall clearance below the platform:

$$F_p = 7.15 - (-1.0) - 1.35 = \underline{6.8 \text{ m}}$$
– confirming the longer fall because anchor is below D-ring.

Free fall distance, $f = 3.5 - (-1.0) = \underline{4.5 \text{ m.}}$

7.5.5 EXAMPLE 7.2

If in the above case, the free fall has to be limited to 1.8 m according to Code, what can be changed and by how much?

Answer:

Either lanyard length L or D-ring distance D may be changed.

To limit the fall, anchor must be above the D-ring, that is, D is positive.

(a) Change lanyard length:

We need, $1.8 = L-(1.0)$, hence, $L = 1.8 + 1.0 = \underline{2.8 \text{ m}}$.

Then, $F_C = 2.8 + 1.25 + 0.05 + 1.35 + 1 = \underline{6.45 \text{ m}}$
$F_P = 6.45 - (1.0) - 1.35 = \underline{4.1 \text{ m.}}$

The price we will be paying for the shorter lanyard length will be a reduced work area.

(b) Change distance of anchor from the D-ring:

We need, $1.8 = 3.5 - D$, hence, $D = 3.5 - 1.8 = \underline{1.7 \text{ m.}}$
As the platform level or D-ring position may not be easy to change, change in D may be effected by raising the anchor by 0.7 m.

$F_C = 3.5 + 1.25 + 0.05 + 1.35 + 1 = \underline{7.15 \text{ m.}}$
$F_P = 7.15 - 1.7 - 1.35 = \underline{4.1 \text{ m}}$, same as for changing lanyard length.

7.5.6 ANCHOR CAPACITY AND NUMBERS

(a) Anchor capacity

The fall arrest force will be quite large, and different codes take different approaches to estimating the adequate anchor capacity.

Often, the harness and other accessories for fall arrest are required to have higher capacity than the anchor itself, because the former are subject to more variable wear and tear.

The USA was the earliest to tackle the question of anchor capacity, and it came up with the big round number of 5,000 lb (22.2 kN) – the popular myth for its choice being that it was the weight of the truck with which the fall impact could be compared. Canada uses the same value.

This recommendation has survived for decades, although Americans are now considering raising it to allow for their increased weight limit of 310 lb (141 kg) for workers and tools, as the average weight of Americans has increased significantly in the last couple of decades.

Singapore has adopted the other extreme, a low figure of 12 kN, on the basis that the limiting force on humans without permanent injury was considered to be 6 kN, and a factor of safety of 2 would be appropriate.

Using the American experience as a proven benchmark, noting that Asians weigh about 70% of European, and fall arrest force being a function of mass, one criterion for fixing the anchor capacity for non-Americans and others equally large would be at 70% of 22.2, that is, 15.4, rounded to 15 kN. Most European and certain other countries provide a capacity of 15 kN.

Table 7.1 lists anchor and related capacities in selected countries.

The fact of the matter is that just as the fall injury cannot be correlated with the fall height, the precise anchor capacity for a particular height of fall also cannot be

TABLE 7.1

Anchor and other capacities in various countries

Country (Code)	Harness & Accessories Capacity, kN	Anchor Capacity, kN
AU (AS/NZS1891)	–	15 kN
Canada (CCOHS)	–	22.2 kN (5,000 lb)
India-BIS (3521-1	15–20 kN	15 kN
IRATA (EU)	22 kN	15 kN
Malaysia	21 kN	15 kN
Singapore (SS528, ISO10333–1)	15 kN	12 kN
UAE	15 kN	15 kN
UK (BS-Ref. 361)	16 kN	15 kN
USA-OSHA (ANSI/ASSE Z359)		22.2 kN (5000 lb)
Average	18.1 kN	15.7 kN

logically determined. The only confirmation for the aptness of the anchor capacity chosen has been the survival rate among the harness wearers that fall.

Recognising the lack of predictability in fall arrest forces, codes such as OSHA permit an alternative capacity of twice the expected maximum arresting force, but this needs to be determined and certified by a competent person.

Going by the average values, and without further research, at least countries with average worker weight in the 70–100 kg range should find the 15 kN anchor capacity quite satisfactory.

Care must be taken to ensure the structural component or other medium into which the anchor is set up can also safely take the specified anchor load. It would not do if the anchor is strong but what it is attached to or embedded into fails!

(b)   Number of anchors

Whatever the anchor capacity requirement is, a sufficient number of anchors must be provided at appropriate locations to facilitate erection and dismantling of temporary structures, and to negotiate gaps and obstacles in edge protection, to satisfy the 100% tie-off rule as described in an earlier chapter.

If, for instance, the person is given a twin-tail lanyard of 1.5 m length, then the span between the two end-clamps of the pair is 3 m – ignoring any gap between the two connecting points at the user's waist. The user should be able to negotiate a gap of 3 m in the edge protection with a single anchor in the middle of the gap, as he can clamp one of the pair to the first point, walk to the second point, anchor the second clamp there, walk back to the first point, unclamp and tuck the end into his waist band, and proceed (see Figure 4.3).

Hence, the number of anchors to negotiate a gap of L m would be $L/(2x)$, where x is the length of twin-tail lanyard, rounded up to integer value, spaced at not more than $2x$.

7.5.7 CHECK FOR OTHER HAZARDS DURING FALL

Once it is decided that there is no alternative to individual fall arrest with a safety harness, and the other corequisites to safety harness described earlier have been implemented satisfactorily, the person in charge of the work at height operation, and the wearer himself if possible, should examine the work site to confirm that:

(a) There are no obstacles in the path of any potential fall;
(b) The harness lanyard is not so long that the fall would catch on an edge and:
   (i) Initiate a swing of the user in a 'pendulum effect';
   (ii) Swing to crash him against a vertical object ('swing back') or smash him onto a horizontal area ('swing down');
   (iii) Slide along a sharp edge and get sliced off, dropping the person.

If any such hazards are found, they should be removed from the path or otherwise immobilised from harming the user.

7.6 USE AND MISUSE OF SAFETY HARNESS

7.6.1 WHEN TO GIVE SAFETY HARNESS

As has been repeatedly emphasised, while the full-body safety harness is a life-saver with its safer redistribution of impact forces and its provision of stopping distance, it carries with it many hazards which demand additional corequisite criteria to be fulfilled.

So, it must be issued with care and its use must be closely controlled.

The most important point that must be remembered is that it is a PPE, which is universally acknowledged as a last resort to be used only when all other options have been considered and found to be infeasible or impossible.

Many codes of practice, therefore, mandate its use mainly where routine procedures expose the workers and other personnel to fall risk without the other safeguards, or while the better safeguards are being implemented, namely:

- Erection of temporary structures like scaffolds and formwork, with their own edge protection or other fall prevention or arrest measures;
- Dismantling of the temporary structures as above after their function has been fulfilled;
- On sloping surfaces such as roofs where other safeguards would not be cost-effective or practical;
- On unstable and moving work platforms such as suspended scaffolds, and mast towers;
- In any other situation with unusual and unknown fall risks.

All of the above are subject to the compliance of all prerequisites mentioned earlier.

An interesting development subsequent to a blanket requirement of safety harnesses for all movable platforms has been the removal of that stipulation for situations where the height is likely to vary from zero to a maximum, and where the occupants already have good edge protection.

A prime example of this is the mobile elevated work platform (MEWP) such as scissor hoist or boom hoist, where obviously safety harnesses may not meet fall clearance and anchor requirements.

As stated in Section 3.4.5, current protection for their occupants is a work restraint by a short lanyard from the waist belt to one of the anchor rings provided at the base of the platform basket with guardrails all around.

Some codes permit users to use safety harnesses in MEWP after they have risen beyond the required fall clearance, or if their task is clearly above it. But this option should be offered with caution where the users are likely to be less knowledgeable or disciplined than would be needed to make the decisions involved.

7.6.2 WHY NOT TO GIVE SAFETY HARNESS INDISCRIMINATELY

We may now summarise the reasons for not giving safety harnesses for all work at height:

- As PPE, it is the least effective control to be chosen after all other controls have been explored and exhausted;
- It is quite costly and being PPE, as many sets will be needed as number of workers involved;
- It must fit the user, and worn correctly; hence, it must be stocked in various sizes, and the correct size must be issued, with their correct wearing ensured;
- It needs more training, better maintenance, and more supervision, than other PPE;
- It needs well-designed anchors of the right capacity and in adequate numbers to provide 100% tie-off;
- It requires a minimum fall clearance calculated for each situation;
- It needs a prompt and proper (preferably on-site) rescue system, to overcome potential suspension trauma;
- It can make the all-day wearer very uncomfortable causing him to make mistakes and/or reducing productivity;
- It may introduce fresh hazards like hitting objects during fall.

Safety harnesses do save lives and are indispensable in many situations where their need has been confirmed and all corequisites are provided, but otherwise, they will pose additional hazards.

7.7 SELF-RETRACTING DEVICES

7.7.1 SELF-RETRACTING LIFELINE

The self-retracting lifeline (SRL), classified as a fall arrest device, works on the principle very similar to that of a seat belt of an automobile. The latter allows the user to draw the belt as far as he needs to strap himself in the seat, but if he jerks on it in will stop; so if the car brakes too suddenly, his body will be forcibly retained within a very short movement forward only to the extent of the slack in the belt, avoiding injury.

Similarly, SRL that consists of a reel with a long cable spooled inside a casing unwinds when pulled as the user moves around doing his task, but if he slips over an unprotected edge or from a ladder rung, a spring mechanism inside activates a brake in response to the jerk, to stop further release of the spooled cable, and stops the fall before the person crashes to the ground or floor.

Unlike the safety harness, the SRL does not allow the user to fall too far, but activates the braking mechanism to a full stop well within 1 m (about 3 ft).

Figure 7.3(a) shows an SRL used by a worker at an unprotected edge high above the ground, with his lanyard clamped to the SRL along an inclined lifeline. In the early days, SRLs were used with lanyards vertical or inclined not more than 15° to the vertical as in the figure.

Recent models have overcome that limitation, enabling workers to use SRLs horizontally also; Figure 7.3(b) shows the SRL used by a worker on a sloping roof, attached to a fixed anchor A from his D-ring D.

The user may need to wear only a waist belt, or a chest harness, as he is not going to fall by more than about 30–60 cm (1–2 ft).

However, some codes require a full-body safety harness for SRL use – without the shock absorber – a main advantage claimed being that the thigh straps of the full-body harness would keep the user from slipping out of the chest harness during or at the end of the fall.

In any case, shock absorber shall not be used with SRL as it will interfere with the spring mechanism inside the SRL designed to stop the fall and possibly cancel its functioning.

Unlike the manual rope grab, SRL does not depend on the worker for the extending or stopping of the cable to reach his work location.

As SRLs face many of the use hazards of safety harnesses, care should be taken that in case of a fall, there is no swing down or swing back, and the cable does not rub against any sharp edge which will cut the cable.

FIGURE 7.3 Self-retracting lifelines.

(a) Original version of SRL.
(b) Current version of SRL.
(c) Personal fall limiter.

Anchor capacity for SRLs is commonly set at 3,000 lb (13.3 kN) with the stipulation that the free fall is restricted to 6 ft (1.8 m); this may be increased to 5,000 lb (22.2 kN) for longer free falls like swing-back or swing-down.

Recent models of SRLs also have a retracting and rescue mode in which a worker who may be unable to function on his own say due to inhaling toxic gas in a confined space may be reeled in to safety by a handle on the casing.

7.7.2 FALL CLEARANCE FOR SRL

Since SRL does not have a shock absorber, and its deceleration distance (a), which will also be its free-fall distance, f is also quite short – around 30 cm (1 ft) – the fall clearance needed will be less than that required for the full-body safety harness.

Note that the lanyard length and D-ring location are no more relevant as the cable will always be taut; hence,

$$F_C = a + h + m. \tag{7.2}$$

7.7.3 EXAMPLE 7.3

A 1.75 m-tall worker has his SRL D-ring 45 cm below the top of his head. The arrest activation distance is 0.4 m. What fall clearance would he need?

Answer:

We have: $a = 0.4$ m, $h = 1.75 - 0.45 = 1.3$ m, and, $m = 1$ m.
By Eq. (7.2), fall clearance required, $C = 0.4 + 1.3 + 1 = \underline{2.7\ m}$.

7.7.4 PERSONAL FALL LIMITER

Self-retracting lifeline models are now available, which may be worn at the back of the user, and the other end of the lanyard has a clamp which the user may attach to any convenient anchor.

This mode, called a 'personal fall limiter' (PFL), a smaller and lighter version of SRL, is specially designed to be worn by individual workers, giving them greater freedom of movement than regular SRLs.

Figure 7.3(c) shows a worker wearing a PFL, the PFL attached at the worker's D-ring, and the lanyard's free end will be attached to the anchor, the opposite of SRL shown earlier.

7.8 FALL PROTECTION PPE

7.8.1 NEW SAFETY HELMET

Although a last resort as PPE, the humble hard hat plays a big role in protecting the head from major injury at the end of falls at the workplace, starting from the 1930s when engineer Strauss of the Golden Gate Bridge came up with his leather hat.

As long as the chin strap, or its modern alternate the band clamp screw at the back, is in place, the hard hat will stay on the head and provide a barrier to any direct impact on the top or side of the head, which is the most vulnerable part of the human anatomy in any fall from height or at level. (The screw clamp, resulting in a constant albeit small pressure around the head, may have its own hazards if applied for long hours!)

Predictably, continuing research into better hard hats has recently come up with the 'safety helmet', with the re-definition of the familiar term referring to the extra padding around the sides and suspension straps across the crown of the head, designed specifically to respond to rotational trauma to the head upon impact of the head with a flat surface [Ref. **7.3**]. It also has a shorter brim in front, to facilitate upward vision.

7.8.2 WEARABLE AIRBAGS

A recent PPE development is the wearable airbag, referring to a set of air sacs worn as a vest around the torso and reaching above the neck at the back, as shown in Figure 7.4 [Ref. **7.4**].

Chip-based micro electro-mechanical system (MEMS) sensors are built into the airbags to detect fall-indicative movements of the wearer, and when they cross preset limits, activate the airbags, much like automobile airbags, in less than a tenth of a second, so that the fall is cushioned at vulnerable locations on the body.

Currently, MEMS is still in research and prototype stage, but with some commercial models being already available, experience with seniors who tend to lose their balance during walking proves promising.

Figure 7.4 displays a wearable airbag product by SAF-T Vest to protect seniors when they fall. Similar principle and product may be used to protect workers from injury when they fall.

While with falling costs, the wearable airbag may become affordable enough to be used to prevent fall injuries at the workplace, the bulky and heavy vest may impede the quality and quantity of work, which may militate against its use.

FIGURE 7.4 Wearable airbag for seniors.

Source and Author: Open Press, SAF-T Vest, anonymised. Link: www.openpr.com/news/ 3008840/wearable-airbag-market-is-booming-worldwide-ducati-motor

7.9   EXERCISES FOR CHAPTER 7

7.9.1   EXERCISE 7.1

A worker 1.75 m tall has his D-ring 40 cm below the top of his head, with a slack of 5 cm. He uses a lanyard 2.5 m long attached to an anchor 0.5 m above the platform level. The shock absorber of his safety harness stretches to 1.2 m. What fall clearance would he need? What will be the free fall distance?

Answer:

Minimum fall clearance below anchor, F_c = <u>6.15 m</u>
Minimum fall clearance below platform, F_p = <u>7.05 m</u>
Free fall distance: f = <u>3.4 m.</u>

7.9.2   EXERCISE 7.2

If in Exercise 7.1, we have only 3 m available for fall clearance and we are willing to settle for 0.5 safety margin, at what minimum height above the platform should the anchor be? What is the free fall height? (<u>Hint:</u> Height of anchor above platform = $D+h$.)

Answer:

Height of anchor above platform, F_p = <u>2.65 m.</u>
   Free fall height, f = <u>1.25 m.</u>

7.9.3   EXERCISE 7.3

A worker 1.7 m tall has his SRL anchored 3 m above his D-ring, which is 1.3 m above his feet, and its activation distance is 0.5 m. What minimum fall clearance would he need?

Answer:

Minimum fall clearance required, <u>F = 2.8 m.</u>

REFERENCES FOR CHAPTER 7

7.1   Krishnamurthy, N., "Full-body safety harness – Blessing or bane?", *The Singapore Engineer – The Magazine of the Institution of Engineers*, Singapore, August 2012, pp. 18–22. Retrievable from: www.profkrishna.com/ProfK-Publications/NK-IES-SE-Aug2012-Harness.pdf

7.2   ———. "Suspension trauma 101", *Blog*, Canadian Safety Group Inc., 18 March 2016. Retrieved on 6 March 2023 from: www.canadiansafetygroup.com/blog/title/suspension-trauma-101/id/13/

7.3   Hall, R., "Safety helmets vs. hard hats", *Construction Dive*, 19 July 2021. Retrieved on 3 May 2023 from: www.constructiondive.com/spons/safety-helmets-vs-hard-hats/603219/

7.4   ———. "SAF-T Vest™ fall injury protection", Retrieved on 7 March 2023 from: https://davenportsaftsystems.com/, and also: www.illustrationx.com/news/4142/the_personal_airbag

8 Lifelines for fall devices

8.1 LIFELINES IN GENERAL

While at height, unless the person is provided with edge protection, he will be tethered by a lanyard, from a waist belt or some kind of harness, to a fall restraint anchor or a fall arrest anchor. This anchor may be either fixed or, for greater mobility, movable along another rope or cable, which in turn will be connected to one or more fixed anchors.

The rope or cable to which the anchor end of lanyards may be connected is called a 'lifeline', a very apt term because the lives of its users literally hang from it! Lifelines are used to access and work vertically at different levels, horizontally at different locations in horizontal or inclined planes.

The design, implementation, and use of lifelines are highly important to the safety and welfare of all users of fall management PPE and other devices and, hence, deserve utmost care.

Although lifelines are a staple item in mountaineering and rock-climbing, the specialised nature of their application in these sport activities throws their treatment beyond the scope of the book. A very clear comparison of the sport lifelines and workplace lifelines may be found in Ref. **8.1**.

Lifelines become necessary when the lanyard is short and the work zone is large,

- if there are too many fixed anchors to provide 100% tie-off for covering the work area, or
- if the user has a rope grab to vary his distance from the anchor.

Even if it were possible to provide many anchors, the consequent need to frequently change anchors would take considerable time and effort, which in turn would not only affect productivity but also potentially create fresh hazards, not all of which might have been identified in the risk management process.

That is why, most of the time in real-life projects of any size, lifelines are used to attach sliding clamps or the lanyard anchors themselves.

8.2 LIFELINES FOR FALL PREVENTION

8.2.1 GENERAL CONSIDERATIONS

As lifelines will be common to all users, their treatment falls under collective fall management, requiring the organisation to plan and implement them properly before the individuals using the lifelines can be expected to keep their side of the bargain in regard to safety.

The lifelines themselves are common to fall prevention and fall arrest, the only difference being in the design capacity, with the former being the less demanding of the two.

Lifeline design depends on many diverse factors all of which cannot be easily quantified. So, countries have widely varying values for the anchor capacity and lifeline strength. There are also available different methods of lifeline design.

In all cases, lifelines must be designed for fall arrest, even if the intent is fall restraint.

8.2.2 TYPES AND APPLICATIONS

Each lifeline will have some cable or rope to which the user will clamp his lanyard or other attachment. Lifelines are usually of synthetic textile material, polyester, polyamide, or nylon, often sheathed (then referred to as 'Kernmantle') in various ways to suit different purposes such as static or dynamic, temporary or permanent, etc.

The rope is usually of polyester or Nylon, of 8–15 mm diameter, with capacities from 12 kN to more than 30 kN. They must be checked for wear and tear before every use.

For the purpose of this book, long vertical metal rail clamps, that is, grooved tracks attached to ladders, in which a sliding element can be held firmly in place at any desired location, will also be referred to as vertical 'lifelines'.

All lifelines must be capable of resisting the maximum impact force due to fall arrest, not less than the specified safety harness anchor capacity.

In recent decades, in place of flexible horizontal lifelines, horizontal bars with sliding rings and 'rigid lifelines' in the form of metal trusses have been used for greater efficiency, economy, and safety.

Thus, there are broadly two types of lifelines for fall control:

1. Vertical lifeline, (a) Textile, and (b) metal track;
2. Horizontal lifeline, (a) Flexible, and, (b) rigid.

8.3   VERTICAL LIFELINES

Vertical lifelines are ropes or cables which hang vertically from an overhead anchor. They are quite easy to design and simple to provide and use. Main uses are for ladders, work positioning, and other situations where vertical movement or positioning is involved.

Needless to reiterate, vertical lifelines are commonly used for rope access and fall rescue.

A vertical lifeline can be used by more than one person, but (in practice) by no more than four. The anchor capacity is expected to be as many times the value for one person as are connected to the same vertical lifeline. Singapore doubles the single person fall arrest value of 12–24 kN for the second person, but limits it only to 26 and 28 kN for the third and fourth persons; this is fairly standard.

The one major hazard to watch out for with vertical lifelines is that when the user moves around beyond about 15° from the vertical, if he happens to lose his footing and drops from his support, the potential injury from swing-back will increase and must be factored into his safety management.

8.3.1 FOR LADDER USE

Vertical lifelines are commonly used as part of fall prevention or arrest system for workers on ladders. To offer the user control of his vertical location and/or horizontal reach, a 'rope grab' is provided for the user to move upward or downward and stop where desired.

The rope is usually of polyester or Nylon, of about 15 mm diameter, and must be checked for wear and tear before every use.

Figure 8.1(a) depicts a vertical lifeline in use by a worker carrying an object; A is the anchor, R the rope grab, S the shock absorber in the safety harness of the user's lanyard.

A similar safeguard, somewhat sturdier than the rope grab, is offered by the 'rail clamp' attached to vertical ladders, Figure 4.5(d), which consists of a grooved rail within which a sliding catch (called a 'trolley' in the West), can be actuated by the user at any desired level, as he climbs up and down to carry out his assigned tasks.

In both cases, the user ring of the slider should be attached to his waist belt or chest harness. Except when climbing up and down, the user has both his hands free and, if on a ladder, he can also lean a little backwards to steady himself while working.

Many codes require the use of a full-body safety harness with rope or rail grabs, for all fall management, in which case, of course, they should be complied with. Otherwise, if they are used for fall restraint as fall prevention, there should be no falls with proper planning and implementation, and hence the safety harness with shock absorber may not be necessary.

The rope, cable, or track should have a stop at the bottom end to prevent the sliding device from slipping out and dropping the user.

FIGURE 8.1   Lifelines for fall management.

(a) Vertical lifeline. [Source and Author: Own work.]

(b) Horizontal lifeline, and shock absorber. [Source and Author: Karam Safety Private Limited; Video screenshot, used with permission. Link: www.youtube.com/watch?v=2x9H-fk9zDq8]

(c) Horizontal lifeline, on multiple supports. [Source and Author: Own work.]

8.3.2 FOR WORK POSITIONING

When the vertical lifeline is used as a fall restraint, that is to say, for work positioning controlled by winches, as there will be no fall, we need to provide anchors and lifelines capable of taking only the maximum load of user, tools, and accessories, times a factor of safety of 2, preferably more, which would be about 4 kN; 1,000 lb (4.5 kN) is considered adequate for vertical work restraint.

As vertical movement may involve some dynamic effects, and to allow for other contingencies, many codes conservatively specify higher figures. America's Occupational Safety and Health Administration (OSHA), assuming a user weight range of 130–310 lb (59–141 kg) requires a capacity of 3,000 lb (13.3 kN) when free fall is limited to 2 ft (0.6 m), and 5,000 lb (22.2 kN) beyond that.

8.3.3 FOR SUSPENDED SCAFFOLDS

In suspended scaffolds, conditions are different from work positioning of single persons. Generally, the scaffold is much heavier than the sit harness that a single person uses; there are also two or more persons at one time moving around on it.

Further, due to the movements of the scaffold and its occupants as well as from wind gusts, the scaffold may tilt or swing about any of the three axes, making the environment quite unstable. That is why safety harnesses are mandatory in most countries for suspended scaffolds.

Hence, lifelines for suspended scaffolds, to be consistent with the requirement of safety harnesses, have to follow the fall arrest criterion such as of 5,000 lb (22.2 kN) or twice the maximum anticipated impact force. In addition, a work restraint lanyard connecting the person's waist belt to a ring at the base of the scaffold is used.

8.4 HORIZONTAL LIFELINES

These are not exactly 'horizontal' lifelines (abbreviated to HLL) as it is well-nigh impossible to get a cable really horizontal, eliminating the natural sag of the line due its self-weight. They are actually 'catenary cables', but we will continue to call them horizontal lifelines.

A horizontal lifeline for a horizontal or slightly inclined slope obviates the possibility of swing falls from a vertical lifeline – with consequent injury potential – as the user moves horizontally farther away from the anchor during his work.

Horizontal lifelines may appear to be safer than vertical lifelines, because horizontal movement should not normally include falls. In the horizontal plane or for small slopes, the lifeline may be used for work restraint, not involving a fall.

However, HLL design and use for fall arrest are a lot more complex than for vertical lifelines. We will address the full range of uses of horizontal lifelines.

8.4.1 FOR WORK RESTRAINT

Horizontal lifelines are very common to facilitate work restraint. In fact, the only way a user can move around outside a circle defined with a fixed location as anchor and a short lanyard length as radius is to provide a horizontal lifeline or bar along which a rope-grab or ring can slide. This topic has been treated in detail in Section 4.3.

Horizontal lifelines are used as work restraint to enable people to traverse a long platform with no edge restraint on one side and as a tether from the ridge of a sloping roof – for gently sloping roofs; as for steeper slopes, the likelihood of rolling down the slope will increase.

(a)   Horizontal plane work

Unlike vertical lifelines which by gravity will automatically take up a straight configuration between the anchor and the user when taut, horizontal lifelines will sag due to their self-weight and/or deflect horizontally due to the pull at the attachment to the user.

These changes must be allowed for while determining the length of lanyard so that the user does not get into a situation where, due to the extra play from the sag, he will face a hazard of falling over the edge while pulling on his tether.

A common value for anchor capacity of horizontal lifelines used as fall restraint is 3.5–4.5 kN (800–1,000 lb).

Some prefer to go up to 6 kN (1,350 lb). Three to four times the weight of the person plus tools and materials carried is also cited for design capacity.

Figure 8.1(b) depicts a worker walking on top of a container with his lanyard attached to an SRL (marked 'S') anchored at A to an HLL supported on brackets. Inset shows the shock absorber E, which is an essential element in HLLs as the impact effect on them from fall arrest forces will be quite high.

Obviously, in both cases above, the lifelines must be designed for fall arrest rather than for fall restraint.

To avoid excessive deflection in a long HLL, such as AG in Figure 8.1(c), intermediate rings on floors, walls, or posts may be provided at intervals, as at B, C, E, and F. These intermediate supports will also facilitate the change of direction of the lifeline in the surfaces of travel and work. The user's lifeline is attached to the HLL at D, which is pulled to the location H if he falls.

The supports may be set up at convenient locations so as not to interfere with the taut lifeline and must have sufficient capacity to resist the maximum estimated forces on them due to the vertical drop, with a factor of safety of 2 or more.

Special procedures must be followed when the user wishes to move from one segment to the next, past a ring or post, to adhere to the 100% tie-off rule already described.

However, for a continuous lifeline looped through intermediate posts, when more than one person attach their anchors to the same line, the pull exerted by anyone in the group will be felt by the others and interfere with their work.

Likewise, a slip or surface fall of one likewise can affect the others. All credible worst-case scenarios for these must be addressed.

(b)   Sloping surface work

For persons on sloping surfaces such as roofs, gravity becomes an unavoidable hazard. Accidental loss of balance and rolling on the surface to the end of the lanyard will induce a jerk upon stopping. These would have to be treated in the same fashion as vertical fall arrest.

(c) Vertical fall arrest

Although the horizontal lifeline may be erected on a horizontal surface, whenever there is a chance of the person attached to a horizontal lifeline falling vertically, it shall be treated as if he is dropping from a horizontal lifeline overhead, as discussed in the next section.

    If the HLL continues over more than one segment, the fall clearance required should include, in addition to the terms in Eq. (7.1), the following:

(i)   Horizontal pull plus vertical sag in the lifeline due to the vertical drop of the falling person (DH in Figure 8.1(c);
(ii) Stretch of the lifeline ABCHEFG due to tension in it;
(iii) Deflections of the supports along which the lifeline passes [Ref. **8.2**].

The other additional concern in lifelines and fall lanyards going beyond the horizontal or sloping surfaces to vertical drops is the distinct possibility that any sharp edge or burr along the edge, as at J in Figure 8.1(c), may damage the rope or cable and even abrade it enough to snap and lead to a fatal crash of the suspended person to the ground or other base.

8.4.2   FOR FALL ARREST FROM HLL

Attaching fall arrest lanyards to flexible horizontal lifelines – as against rigid lifelines – should be avoided if at all possible. The reason is that when a taut horizontal lifeline is subjected to a vertical load, even the smallest deflection will cause a highly magnified horizontal pull at the two ends.

    A simple resolution of forces will confirm that if a lateral pull of 1 kg on a taut (weightless) horizontal cable 1 m long deflects it by 1 mm, the end reaction is (500 mm/1 mm) or 500 times 1 kg, that is 500 kg! As cables will naturally sag due to their self-weight, the additional sag due to the person's fall will not produce such a catastrophic end reaction, but still, the pull on the supports will be quite large.

8.5   DESIGN CONSIDERATIONS FOR FLEXIBLE HLL

8.5.1   INITIAL CABLE SAG AND TENSION

The shape of a draped cable weighing w per unit length will be a catenary, and the relationship between sag (f_1) and the tension (T_1) at the ends on a span (L) is given by

$$f_1 = (L/4)/\sqrt{\left\{\left[(2T_1)/(wL)\right]^2 - 1\right\}} \tag{8.1a}$$

$$\text{or, } T_1 = (wL/2)\sqrt{\left\{\left[L^2/(16f_1^2)\right] + 1\right\}}. \tag{8.1b}$$

Without some tension or sufficient overhanging cable at the two ends, the HLL cable will simply slip off the supports under its own weight. Even with tension, there will be some sag; greater the tension, less the sag.

Too much tension will impose undue stresses on the lifeline and the supports, demanding larger and heavier cables and supports. Too little tension will increase the sag, increasing the free fall distance, which in turn will demand larger fall clearance.

The initial sag and tension should, therefore, be carefully chosen and implemented.

Proprietary systems recommend initial sags for different spans of HLL; they are generally less than 1% of the span. Ref. **8.3** gives such a chart for up to 82 ft (25 m), which may be represented by the following equations:

In feet: $f_1 = 0.0001L^2 - 0.0011L + 0.0808$ (8.2a)

And, in metres: $f_1 = 0.0003L^2 - 0.0011L + 0.0246.$ (8.2b)

Beyond 25 m, we may limit sag to less than 1%.

Note that Equations (8.2a) and (8.3b) are the same curve in the different units.

Once the sag is chosen, the initial tension to get the sag may be found from Eq. (8.1b).

8.5.2 EXAMPLE 8.1

What will be the initial sag for an HLL with initial tension of 30 kg, for a cable of 10 m span, weighing 0.25 kg/m?

Answer:

By Eq. (8.1), the initial sag is:

$f_1 = (10/4)/\sqrt{\{[(2 \times 30)/(0.25 \times 10)]^2 - 1\}} = \underline{0.104 \text{ m.}}$

8.5.3 EXAMPLE 8.2

What is the initial tension required for a span of 20 ft using a 16 mm polyester cable weighing 0.179 kg/m?

Answer:

From Eq. (8.2a), the recommended sag for a span of 20 ft (6.1 m) is:

$f_1 = 0.0001 \times 20^2 - 0.0011 \times 20 + 0.0808 = 0.099 \text{ ft} = 1.19 \text{ in} = 0.030 \text{ m.}$

From Eq. (8.2b), the recommended sag for 6.1 m span is:

$f_1 = 0.0003 \times 6.1^2 - 0.0011 \times 6.1 + 0.0246 = 0.029 \text{ m, very close.}$

The cable weighs 0.179 kg/m = 0.178 × 9.81 = 1.76 N/m.

The initial tension, by Eq. (8.1b), working with metric units, is:

$$T_1 = \{\sqrt{[6.1^2/(16 \times 0.03^2)+1]}\} \times 1.76 \times 6.1/2 = 272.9 \text{ N}$$
$$= 272.9/9.81, \text{ that is, } \underline{27.82 \text{ kg}}.$$

In Imperial units, the initial tension $= 27.82 \times 2.205 = \underline{61.3 \text{ lb}}$.

8.5.4 FALL ARREST SAG AND TENSION

(a) HLL analysis

The analysis of what happens after a person falls with a safety harness attached to an HLL is quite complicated.

The two detailed applications of ANSI/ASSP's Z359 techniques for HLL design given in Ref. **8.4** will illustrate the many considerations and numerous variables involved.

Just to savour the flavour of this difficult procedure, we will go through one of the simpler expositions [Ref. **8.5**] of the design process, with reference to Figure 8.2, which gives the sketch of and governing equations for a single span HLL, omitting the contribution of the deflection of the posts.

Note that Eqs. (i-a) and (i-b) in Figure 8.2 are alternative formulations of the same relationship, to find either f_1 or T_1, given the other.

The single main consideration in HLL design and implementation is that the user shall not be subject to more than the permissible arrest force F, broadly accepted as 4 or 6 kN, occasionally relaxed to 8 or 10 kN.

(b) HLL design procedure

(1) Find f_1 or T_1 from Eq. (i-a) or (i-b) of Figure 8.2, depending on whether we know (or assume) T_1 or f_1.

(2) Calculate s_1 from Eq. (ii) of Figure 8.2.

(3) To find $\cos(\alpha)$, we need T_2, which in turn involves $\cos(\alpha)$. Hence $\cos(\alpha)$ and T_2 must be solved for iteratively – this is best done through a spreadsheet program – as follows:

  (i) Assume some initial tension T_2 (say $= 2F$), calculate $\cos(\alpha)$ from Eq. (iii).

  (ii) Find T_2 from Eq. (iv), using the chosen value of F.

  (iii) Use this new value of T_2 to find revised value of $\cos(\alpha)$ from Eq. (iii). Repeat steps (ii) and (iii) until satisfactory convergence is obtained.

[I usually take a rough average between the new T_2 and the previous T_2, hoping it would take me faster to the right solution, but it should not matter!]

(4) Find the total final sag f_2 from Eq. (v) of Figure 8.2.

8.5.5 EXAMPLE 8.3

An HLL is a 30 m-long, 9.5 mm-diameter cable weighing 3.65 N/m; its cross-sectional area A is 42 mm² and the modulus of elasticity E is 65 GPa. Initial tension

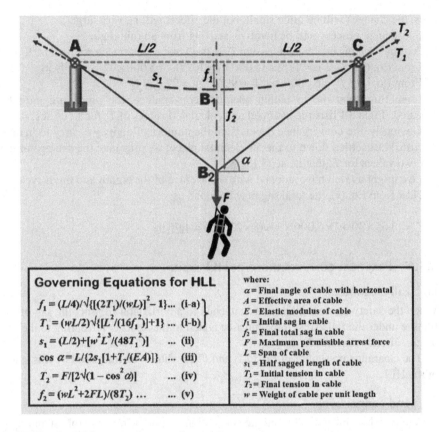

FIGURE 8.2 HLL nomenclature and governing equations.

Source: Ref. **8.5**. Author of Figure: Own work. Link to Ref. **8.5**: www.researchgate.net/publication/350767744_Design_of_Horizontal_Lifeline_Systems_for_Fall_Protection

is 5 kN. If the maximum arrest force F is to be limited to 6 kN, what should be the maximum tension? What will be the total sag?

Answer:

[NOTE: The area of a 9.5 mm-diameter circle is $\pi(9.5^2/4)$, that is, 70.9 mm². However, the effective area of the strands (inside the wrapping) that take the force is known to be 42 mm², amounting to 59.2%. Values for other diameters and numbers of strands may be obtained from manufacturers' tables.]

For the initial tension T_1 of 5000 N, the sag by Eq. (i-a) of Figure 8.2 is:

$$f_1 = (30/4)/\sqrt{\{[(2\times5,000)/(3.65\times30)]^2-1\}} = \underline{0.082 \text{ m.}}$$

From Eq. (ii), initial half-length of sagged cable,

$$s_1 = 30/2 + (3.65^2\times30^3)/(48\times5000^2) = \underline{15.0003 \text{ m}}$$

(Yes, the changes will be quite small, but the effects will be very large!)

Subsequent process will be iterative, starting from an initial guess.

Assume starting estimate of tension T_2 as (say) twice the arrest force, that is, 10 kN.

From Eq. (iii): Cos α=30/{2 × 15.0003[1 + 10,000/(65,000 × 42)]} = 0.9956

From Eq. (iv): T_2 = 6,000/[2×√(1−0.9956^2)] = 32,029 N

From this point on, to obtain adequate convergence, the procedure must be repeated. Table 8.1 lists the assumed and calculated values of T_2 for 12 cycles.

Obviously, the convergence depends on the number of digits we carry to iterate; as cos(α) has settled down to the 5th decimal place, we may take the average of the last two values for T_2, that is, at 23,155 N.

(We might as well have stopped with the average of the eighth and ninth cycles.).

Now from Eq. (v), the total sag may be found as:

$$f_2 = (3.65 \times 30^2 + 2 \times 6,000 \times 30)/(8 \times 23,155) = 1.96 \text{ m}$$

8.5.6   Practical Considerations in HLL Design

(a)   Fall clearance

When the safety harness clamps are attached to a horizontal lifeline, the sag of the lifeline under the fall arrest force must be added to the items contributing to the fall height, already discussed.

The clearance presented in Eq. (7.1) and (7.21) should be increased by this sag f_2 of the HLL.

(b)   End connections

Normally, HLL end and intermediate connections would be made to robust and rigid supports, and hence their deflection contribution will be quite small and may be ignored, unless the connection is to the top of relatively slender posts.

TABLE 8.1

Iteration steps to determine additional tension in cable.

Itern. No.	Assum-ed T_2, N	Cos(α)	Calc. T_2, N
1	10,000	0.9956	32,029
2	32.029	0.9884	19,740
3	19,740	0.9928	25,047
4	25,047	0.9909	22,275
5	22,275	0.9919	23,599
6	23,599	0.9914	22,937
7	22,937	0.9916	23,261
8	23,261	0.9915	23,101
9	23,101	0.99159	23,180
10	23,180	0.99156	23,141
11	23,141	0.99157	23,160
12	23,160	0.99157	23,150

The cable tensions are usually quite high, and most proprietary HLL systems incorporate some kind of energy absorbing device to reduce the size of the cable and its anchor.

The energy absorbers, acting like oil-damped automobile shock absorbers, are usually bending members, metal folded plates, or strong springs, which will elastically extend under force, and gradually recover their original configuration.

End anchors and connecting hardware for HLL are often required to have a high minimum ultimate strength, a typical value being 16,000 lb (71 kN) [Ref. **8.6**], and the cables themselves to be even stronger, at 20,000 lb (89 kN) [Ref. **8.7**].

(c) Current status of HLL design

Except in countries with advanced safety-related technology, a sad commentary on the existing state of affairs in the design of HLL is that in spite of the availability of simplified analysis procedures and even free or modestly priced computer software for them, professional designers too miss certain important factors to be covered in the analysis and end up with flawed design, leading to tragedy and loss at the workplace.

A paper by Wang *et al* [Ref. **8.8**] points out common mistakes occurring in HLL design, potentially leading to grave consequences. The deficiencies they list are:

* Necessary considerations omitted;
* Inappropriate and inconsistent design assumptions used;
* Inappropriate analytical method used;
* Intent of regulations not addressed.

All this is still not the last word on HLL design. There is more, but all that is another story!

It would be safer to leave HLL design to the experts or simply go in for proprietary HLLs (or less expensive alternatives, again proprietary products), for then the burden of safe design for a set of known circumstances would be on the manufacturer and supplier!

8.6 RIGID HORIZONTAL LIFELINES

In recent decades, the fact that while a flexible cable deflects a lot under a load and consequently its design is complex and risky, a truss or beam which deflects very little under the same load has been put to good use for horizontal lifelines.

Rigid lifelines avoid all the complicated calculations and resultant fall sag and tension complications and equipment costs of flexible lifelines. Of course, the rigid lifelines too will involve erection and support considerations, entailing additional costs, and appropriate safety management. Design of rigid HLLs is relatively much simpler than for flexible HLLs, mainly because the truss or beam deflections may be directly calculated, and quite small.

8.7 EXERCISES FOR CHAPTER 8

NOTE: It should have become clear from the steps in the worked examples that if many such problems need to be worked out – and definitely for iterative schemes – it may be more efficient and faster to code the expressions to be calculated into a pro-grammable calculator or spreadsheet than enter the numbers and operators for for-mulas into a calculator. Any coding must be validated by manual solution!

8.7.1   EXERCISE 8.1

What will be the initial sag for an HLL with initial tension of 500 N, for a cable of 20 m span, weighing 3 N/m?

Answer:

Initial sag, $f_1 = 0.301$ m.

8.7.2   EXERCISE 8.2

What initial tension is required for a 15 m span, with a polyester cable weighing 2.5 N/m, if sag must not be more than 10 cm?

Answer:

Initial tension, $T_1 = 703.4$ N.

8.7.3   EXERCISE 8.3

An HLL is 25 m long, the cable weighing 3.5 N/m; its cross-sectional area A is 50 mm^2 and the modulus of elasticity E is 65 GPa. Initial tension is 4 kN. If the max-imum arrest force F is to be limited to 5 kN, what should be the maximum tension? What will be the total sag?

Answer:

Maximum tension, $T_2 = 21.71$ N; Total sag, $f_2 = 1.452$ m.

8.7.4   EXERCISE 8.4

If in Exercise 8.3, the total sag must be limited to 1.2 m due to lack of clearance, what shall be the maximum tension and consequent maximum arrest force?

(Hint: This is one level higher iteration than when F is given. Equation (v) of Figure 8.2 shows change of sag will involve tension T_2, and Eq. (iv) shows change of tension will involve change of arrest force F. Therefore, it would be best to assume a value for F (– smaller for the smaller sag), find the tension and the sag; repeat until the sag is close to 1.2 m.)

Answer:

Max. arrest force, $F = 2.75$ kN. Max. tension, $T_2 = 14.55$ kN.

REFERENCES FOR CHAPTER 8

8.1 ———. "Can I use my rock climbing equipment as industrial fall protection?", *Rigid Lifelines*, 9 August 2013. Retrieved on 13 August 2013 from: www.rigidlifelines.com/blog/can-i-use-my-rock-climbing-equipment-as-industrial-fall-protection/

8.2 ———. "Horizontal lifeline calculation", *XS Platforms*. Retrieved on 10 March 2023 from: https://fallprotectionxs.com/horizontal-lifeline-calculation/

8.3 ———. *Instruction manual for Big Boss HLL*. Guardian Fall Protection, UK. Retrieved on 11 March 2023 from: www.engineeredfallprotection.com/pdf/guardian-big-boss-horizontal-lifeline-manual.pdf

8.4 Hoe, Y.P., and Goh, Y.M., "Designing and calculating for flexible horizontal Lifelines based on design code CSAZ259.16", *The Singapore Engineer*, October 2014.

8.5 Galy, B., and Lan, A., "Horizontal lifelines – Review of regulations and simple design method considering anchorage rigidity", *International Journal of Occupational Safety and Ergonomics*, March 2017. Retrieved on 7 November 2023 from: www.researchgate.net/publication/350767744_Design_of_Horizontal_Lifeline_Systems_for_Fall_Protection

8.6 Ismail, K., "Personal fall protection equipment", *HSSE World*. Retrieved on March 2023 from: https://hsseworld.com/personal-fall-protection-equipment/

8.7 ———. "Fall protection – Legislation for anchor strength", *Canadian Center for Occupational Health and Safety (CCOHS)*. Retrieved on March 2023 from: www.ccohs.ca/oshanswers/hsprograms/fall/fall_protection_leg_anchor_strength.html

8.8 Wang, Q., Hoe, Y.P., and Goh, Y.M., "Evaluating the inadequacies of horizontal lifeline design: Case studies in Singapore", *Proceedings of CIB W099 Achieving Sustainable Construction Health and Safety*, Lund, Sweden, 2–3 June 2014. pp. 660–670.

9 Planning fall management

9.1 CATEGORISATION OF FALLS

In earlier chapters, we have treated falls of all kinds, without distinction of purpose, instead sorting them according to prevention versus arrest, and collective versus individual.

There is another way of viewing falls. If we look at falls which come under the purview of fall management, four trends are clearly seen:

(1) Young children, old people, and persons who are not in full control of their movements fall without intending to, without warning and without being prepared for it. These are purely accidental falls, and very little can be done about it, except employ support personnel to monitor and manage the falls.
(2) Normal able-bodied adults accessing or working at height when exposed to falling from height, surely do not want to fall, falling not being a normal part of their activity. But if falls happen, they want to avoid death and minimise injury; these are accidental planned falls. They and their employer have to plan and implement fall safeguards. Examples are construction and manufacturing;
(3) Normal able-bodied adults engage in sport, competition, entertainment or other activity which directly or indirectly will involve falls. Falling is an unavoidable result of their activity, but they want to avoid death and minimise injury due to the fall; these are undesired planned falls. They and their sponsors must plan and implement fall controls. Examples are pole vault and movie stunts;
(4) Normal able-bodied adults jump to fall on purpose, in an activity in which jumping or falling down is the key event, but understandably, the performers wish to avoid death and minimise injury; these are intentional planned falls, and the jumpers and their supporters must plan and implement fall management. Examples are circus and bungee jump.

The common overarching concern in all of these four categories is the need and desire to prevent loss of life or limb. (Naturally, we exclude suicides, which are intentional falls to do terminal self-harm.)

In this chapter, we shall review the considerations in both fall avoidance and fall planning.

9.2 ACCIDENTAL FALLS

Accidental falls are very common with the very young and the very old; in particular, more than 25% of people older than 65 years fall are injured. Almost every child falls before it can walk.

DOI: 10.1201/9781032648132-9

Falls at construction, fabrication, maintenance, inspection, supervision, and such other work at workplaces and during professional work have a major impact on the workforce safety and employer liability.

The toll taken on worker lives by objects falling on them is also quite high at workplaces; this will be discussed further later.

The global focus on accidental falls, particularly at the workplace, is because the professional responsibility and legal liability for worker falls will land on the employer, and the immediate and collateral costs of such accidents are a huge drain on industry and society. For these reasons, safety standards for fall management have been evolved and enforced by nations.

9.3   HIERARCHY OF FALL MANAGEMENT

The logical sequence of management decisions to be made for work at height is as follows:

(1) If work at height can be avoided, that would be best (Sections 3.1 and 3.2).
(2) If work at height is unavoidable, adopt collective fall prevention by edge protection (Sections 3.3–3.5).
(3) If collective fall prevention is not feasible, adopt individual fall prevention by:
   (a) Fall restraint in horizontal or gently sloping surfaces (Sections 4.1–4.3);
   (b) Work positioning in steeply sloping surfaces and for vertical positioning at desired levels (Sections 4.4–4.6).
(4) If individual fall prevention is not feasible, adopt collective fall arrest by soft landing, interposing above the hard ground or base:
   (a) Safety net (Sections 6.2 and 6.3);
   (b) Cushion, airbag (Sections 6.4 and 6.5).
(5) If collective fall arrest is not feasible, adopt individual fall arrest by:
   (a) Safety harness, satisfying all corequisite criteria (Sections 7.2–7.5);
   (b) Self-Retracting Devices (Section 7.7);
   (c) Wearable airbags (Section 7.8.2).
(6) If falls are unavoidable, provide one or more of the following measures aimed to reduce fall injury:
   (a) Increase stopping distance of fall arrest;
   (b) Use devices and train personnel in presenting a less vulnerable part of the body over as large an area of contact as possible.
(7) If fall restraint or fall arrest extends over a large area, select or design the appropriate lifeline (Chapter 8).

Proper planning should include the following administrative measures also:

(8) Train and conduct daily briefings to the workers on the proper wearing of the PPE, particularly the safety harness;
(9) Continuously supervise all work at height, ensuring the strict implementation of every step in Safe Work Procedures and the proper wearing and use of all PPE;

(10) Prepare and circulate work at height plan – rescue plan figuring prominently – specific to the project;

(11) Train, brief and conduct practice drills on rescue and post-rescue measures in case of arrested falls, as described in the various earlier chapters.

Figure 9.1 displays this holistic hierarchy of fall management at the workplace.

In spite of such clear evidence of the validity of the above hierarchy of fall management measures, it is disheartening that in certain countries, many companies, when they are unable to provide edge protection, jump to the safety harness as the next safety alternative, either not realising, or even knowing but ignoring, the fact that there are at least a couple of proven better options which they could – and according to accepted principles of risk management, should – try first.

9.4 PLANNING FOR WORK AT HEIGHT AND RESCUE

9.4.1 WORK AT HEIGHT PLAN

Components of a typical fall protection plan, according to the Singapore Ministry of Manpower, are as follows [Ref. 9.1].

(a) Policy for fall protection

• Stating the organisation's approach to fall protection, and management commitment for funds and support;

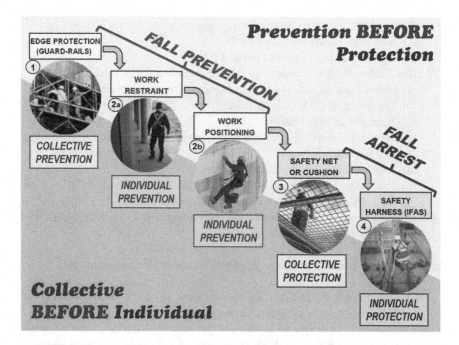

FIGURE 9.1   Holistic hierarchy for workplace fall management.

Source and Author: Own Work, with thumbnails of generic images.

(b) Responsibilities of stakeholders
- From top management down to the worker, for prevention of falls and of injuries due to falling;

(c) Risk assessment and control measures
- Identified, evaluated and implemented, to eliminate fall risks during work, or at least mitigate their consequences;

(d) Fall Protection equipment
- Carefully chosen and issued to users, with adequate training, to reduce risks to acceptable levels;

(e) Inspection and maintenance
- Documented regimen to ensure systems and equipment are in good working order and serve their purpose. For major projects, a 'Permit to Work' detailing the tasks of work at height, hazards therein, measures to prevent mishaps or mitigate their consequences, and certification of the implementation of the proposed safeguards will be essential.

(f) Training
- To identify hazards, assess risks, and understand and observe all necessary control measures, including PPE;

(g) Mishap investigations
- Fall accidents or serious incidents investigated to find deficiencies, and devise corrective/preventive actions;

(h) Emergency preparedness
- Emergency response procedures to be established and implemented, tested, and revised for effectiveness.

A very elaborate fall protection plan is available from the OSHA, as Ref. **9.2**.

9.4.2 FALL RESCUE PLAN

Rescue of persons suspended from safety harnesses has been discussed in Chapter 7 on individual fall arrest. Rescue may be in one of the following forms:

(a) Self-rescue, with the fallen person climbing back to safety on his own.
(b) Assisted self-rescue, with the fallen person being provided with foot stirrups, mechanical winches, or rope ladders.
(c) Fully assisted rescue, by rescue personnel, with rescue equipment or by rope access.

Many standard rescue plan templates are available on the Internet. The National Safety Council of the USA offers detailed guidelines for the creation of a fall rescue plan in Ref. **9.3**.

The Health and Safety Executive (HSE) of the UK states that all working at height rescue plans must address the following:

- The safety of all individuals carrying out the rescue.
- Information about the anchor points for the safety equipment.
- The suitability of any equipment.
- How the individual will be attached to the rescue equipment.
- How the individual will be moved using the rescue equipment.
- Information about any medical needs or other needs of the individuals involved in the rescue.

Any planned rescue equipment must be ready and functional at the worksite prior to commencement of the work at height.

9.5 FREAK SURVIVALS FROM FALLS

Mention must be made for the unexpected survival of a few fall victims, both accidental and intentional (including suicidal), if only because they prove the validity of the physics and dynamics principles presented in Chapter 5.

9.5.1 SURVIVALS FROM ACCIDENTAL FALLS

There have been a few cases of parachute jumpers whose parachute failed to open, but luckily, they fell on glass roofs, bushy wooded areas, or snow-covered terrain.

There are many examples of parachutes failing to open, but the parachutists happened to fall on trees, and the parachute cables snagged on tree branches slowing their descent to life-saving levels.

There are also cases of people falling from airplanes without (or with non-functioning) parachutes and managing to survive. How did they survive?

1942:

Ivan Chisov intentionally (for military strategic reasons) delayed opening his parachute until too late, but luckily fell on a snowy ravine. Possibly the unopened parachute cushioned and spread the load on his back. He endured major injuries but recovered completely.

1943:

Alan Magee, his bomber plane shot down in WWII, his parachute rendered useless, happened to fall on the glass roof of the railway station, Ref. **9.4**. He received many injuries including to his nose and ear, but lived to 82 years of age with his diminished physique.

1944:

Nicholas Alkemade, also a bomber crew member, jumped out of his burning plane having lost his parachute to the flames. His fall was broken by pine trees and a soft snow cover on the ground. He suffered only a sprained leg, and continued to work after the war.

1971:

Juliane Coepcke fell still strapped to her seat in a passenger plane. She suffered a collarbone injury due to the impact. Her survival was probably due to the fact that her middle seat was still attached to the two seats on either side, increasing the air resistance – augmented by an impending storm – and cushioning the impact, which was in a wooded area with thick foliage. In fact, 14 other passengers had also survived likewise, but died during the 11-day long wait for rescue, which Juliane somehow overcame.

1972:

Vesna Vulovic fell trapped within the plane fuselage that broke off due to a terrorist bomb, and the fuselage landed at an angle in a heavily wooded and snow-covered mountainside, which cushioned the impact. She had multiple major fractures and other injuries, but recovered almost completely.

1996:

Bear Grylls, skydiving, fell to the ground on his back when his parachute failed, and lives in pain from herniated discs (spine injury).

2001:

Christine McKenzie fell during skydiving, fracturing her pelvis, but recovered fully; she attributes her light injury to having her fall was broken when she bounced on some powerlines before hitting the ground.

2008:

Chick Walz's fall from inside his damaged balloon was broken by a tree, survived with a broken pelvis and other injuries.

2009:

James Boole, concentrating on his photography assignment as he dropped from the plane, forgot to open his parachute in time, and hit the snow-laden ground hard, breaking his back despite the snow cushion. He returned to reasonably active life after surgery and recuperation.

The common factor in most of these survival cases was the completely unplanned and unexpected availability of increased area for air resistance, or of a cushioning medium such as snow, foliage, own and two side seats, glass roof, and powerlines.

9.5.2 SURVIVALS FROM SUICIDE FALLS

Although I had decided not to discuss suicide falls, two cases beg for attention because the reasons for both survivals are similar, and traceable to sound physical principles.

As a prelude to discussing those cases, it must be reiterated that the fall – or any crash – arrest phenomenon – is basically a matter of the dissipation or transfer of the kinetic energy due to the velocity of the person into some other action ('work done'). If most of this energy transfer can be diverted away from crushing the body, then the person survives.

The most direct and practical way to transfer the kinetic energy is by slowing down over a stopping distance, the medium commonly being a soft landing such as net or cushion, or by the shock absorber of safety harness, coiled spring of SRL, etc., as already described.

However, stiffer materials – like cardboard boxes already mentioned – are also theoretically good shock absorbers, although impact of our bodies on to such harder surfaces will be quite painful and potentially harmful, even when not fatal.

Occasionally, however, metal tops and glass windshields of cars have saved jumpers – who had not planned to end their lives by falling on the cars – from death.

(a)   Man's 39 storey fall

Twenty-two-year-old Thomas Magill plunged 39 storeys from a New York West Side high-rise in September 2010, crashing through the rear windshield of a sports car parked below. He broke both his legs but was otherwise uninjured and recovered normally [Ref. **9.5**].

(b)   Woman's 23 storey fall

Thirty-year-old woman who jumped from the 23rd storey of a Buenos Aires hotel (inset) in January 2011 happened to land on the roof of a parked taxi cab on the street below. She received multiple injuries but recovered quite well [Ref. **9.6**].

Clearly, it was the energy necessary to break the car's rear windshield and crumple its steel trunk cover in the man's case, and break the car's front windshield and bend the taxi's metal roof in the woman's case, that absorbed the kinetic energy of the long drop.

The stopping distances were only a third to half a metre, and hence the G-forces on the two persons would have been very high. Obviously, the bodies withstood such near-instantaneous large Gs with survivable injuries, presumably because both fell on their backs presenting a less vulnerable limb with a large area of contact.

Needless to say, if they had fallen on their heads, the metal impact, the small areas of impact, and the small stopping distance would have spelt their death.

These, and a few other similar cases in many other parts of the world, validate the mechanics principle that transfer of the kinetic energy of fall to the crumpling or crushing of some other object is a viable escape from certain death!

9.6   INTENTIONAL JUMPS FROM HEIGHT

It should be increasingly clear that we can control the outcome of a high fall with proper planning and will to succeed.

Although not for everybody or to be recommended for routine work at heights, the fact that some daring souls have taken jumping safely from height in a pre-planned manner as a challenge, and demonstrated that such a feat was possible, is worth some detailed analysis.

The study may hold some lessons for those aspiring for safety in work at height, at least what to avoid, if not what to emulate.

9.6.1   WHY FALL INTENTIONALLY

As already mentioned, there are situations in modern life, where falling from height is the actual task and not an unwelcome side effect of 'normal' work.

There is the entertainment industry such as film and TV in which falls of ever-increasing difficulty will have to be actually done – usually by stunt-doubles for the heroes and villains who are expected to demonstrate their heroism or villainy for the viewers.

Trapeze, wire walking and cycling, and other high-flying acts in circuses continue to be star attractions for people of all ages.

Moreover, more and more people are seeking and finding safe ways of jumping or voluntarily falling from heights for various reasons including fun activities like paragliding and bungee jumping.

There is also the ever-increasing lure of competitive sports such as the Olympic games, and of adventure fun such as mountaineering, in all of which fall injuries are possible in spite of the care taken by the organisers and the voluntary participants, as already discussed in Chapter 6.

These 'intentional' falls are evidence that safe falls can be planned and executed so that the participants perform unscathed.

Unlike voluntary risk-takers – such as for instance, the parkour enthusiasts who do not care about their life or limb should they fail in their determined but unpractised adventures – intentional fallers take considerable care that they do not hurt themselves.

They spend a lot of time and energy learning and practising their jumps or ascents and descents under expert guidance and with application of appropriate current science and technology.

9.6.2 INTENTIONAL FALLS ON CARDBOARD BOXES

One of the fall protection technologies which found practical use is the use of cardboard boxes for individual fall arrest described in Section 6.5.

Actor doubles who do motor-cycle jumps across ravines and leap over raging fires routinely end their crashes off-camera on cardboard box assemblies.

Many Westerners have done and continue to prove to themselves and the world they can jump from heights with no fear and no injury.

(a) Ferdi Fischer jumping from 45 m

On 11 January 2010, American Stuntman Ferdi Fischer jumped 45 m (150 ft) from a cliff, landing on his back to cardboard boxes demonstrating his courage and the ability of empty cardboard boxes to cushion falls [Ref. **9.7**]. He did a few tumbles in the air before hitting the box stack on his back and walking out from the bottom in minutes.

Taking a conservative guess at the height of the box stack at 4 m, and assuming the entire stack depth was used up, the impact force would be about 45/4 or 11 G, tolerable for the short duration of the stopping.

(b) Gary Connery jumping from 732 m

On 23 May 2012, Gary Connery of UK jumped from a helicopter at 2400 ft (732 m) height, Figure 9.2(a), with a wingsuit (and a back-up parachute, which he did not have to use) and landed on a stack of cardboard boxes, Figure 9.2(b), marked 'X' in both figures. He walked out on his own from the base of the stack, Ref. **9.8**.

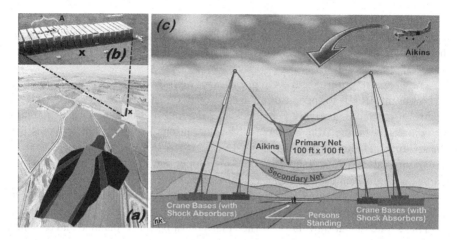

FIGURE 9.2 Cardboard box and net for fall arrest.

Source and author: Own work, based on images from videos, photographs and other material on the internet.
(a) Gary Connery in a wing-suit ready to land on cardboard boxes.
(b) Gary Connery's stack of cardboard boxes.
(c) Luke Aikins, landing in a net after a 25,000 ft fall.

The wingsuit slowed him considerably and gave him good control of his landing trajectory, so the normal height-velocity relationships would not hold.

In the words of Trey West, one of the commentators,

> This is amazing when you think about all the planning that went into it! The physics involved in doing this, . . . the weight of a falling body in a wingsuit, the height of the drop, the wind speed, the direction of the wind, the down draft from the blades of the helicopter, the angle of insertion from the drop moment, the trajectory that is needed, the angle of approach, the flare point, etc.

The video [Ref. **9.9**] shows Connery aligning his final approach to the landing before he crashes on to the box stack. The landing gash A is marked in Figure 9.2(b) with a flower bracket of length somewhat greater than the width, estimated around 12 m, and the depth to the base is about a third of the width in plan, about 4 m, the diagonal distance becoming the 'stopping distance', estimated to be about 13 m.

As for the impact force, Connery is known to have slowed down to about 80 km/h [Ref. **9.8**], that is 22.2 m/s, with his wing-suit at landing, at which speed, the equivalent fall height would be $[22.2^2/(2 \times 9.81)]$, that is, 25.1 m; then the impact would be (25.1/13 m) that is 1.9 G, which is not harmful at all.

The hardness of the cardboard might have scratched or led to some light blunt trauma to any contact surface with the body.

(c)   David Blaine jumping 30 m

In May 2022, famous illusionist and endurance artist David Blaine jumped 30 m (100 ft) on to stacked cardboard boxes, after sitting on top of the pole for 35 hours [Ref. **9.10**].

He dropped on his back into the 3.7 m (12 ft) high stack of boxes, and suffered a mild concussion, probably because of his weakened condition from his long perch atop.

He might have experienced an impact force of (30/3.7) that is 8.1 G, which, spread over the expanse of his back, must have been quite possible to bear.

Just to warn how such jumps are not 'easy' or 'guaranteed safe', Blaine during his Las Vegas (USA) show on 10 March 2023, jumped 'routinely' from 80 ft (24 m), but missed the cardboard box target and dislocated his shoulder, luckily reset by doctors on the spot and thankfully endured nothing worse!

9.6.3 Aikins' intentional fall on a net

This last case study will be a testimony to the ingenuity, heroism, and determination of man against nature!

On 30 July 2016, American professional skydiver and pilot Luke Aikins, then 42 years old, jumped from a plane flying at 25,000 ft (7,620 m) without any wingsuit or parachute and landed without any hitch or harm in a specially designed fall arrest system comprising two nets and four compressed air shock absorbers, Figure 9.2(c).

When Aikins reached the net, he was hurtling downwards at about 120 mph (his terminal velocity), that is, 176 fps (54 m/s), which would theoretically correspond to a fall height of $176^2/(2 \times 32.2)$, or 481 ft (147 m).

The primary net deflection estimated from Figure 9.2(c) was about 50 ft (15.2 m), confirmed by the duration of the stop 1.8 s. Based on this stopping distance alone, the impact force would have been (481/50) or about 9.6 G, which, spread over his back and sides would have been well within his bearable stress levels.

Actually, the compression cylinders at the four poles would have taken a big share of the first impact, and so, based on Aikins' comments later, he appeared to have experienced only 3 G.

The video of the fall [Ref. **9.11**] shows Aikins leaping out of the plane at 25,000 ft, handing over at 15,000 ft his oxygen mask (which he had to use in the rarefied atmosphere of high altitudes), practising flips over to his back at some distance above the 'sweet spot' of the net, and gracefully landing on the soft net.

When the net stops stretching and Aikins is lowered to ground level, he simply walks out, hugs his wife, grabs his 4-year-old son, and gives an interview. This ends the historic, epic event.

This apparently perfect execution of a perfect plan did not come easy, fast, or cheap.

Aikins had assembled a high-powered team of physiologists, professors, scientists, engineers, fall arrest industry leaders, weather experts, skydiving veterans, and event managers nearly a year ahead of the scheduled date and outlined his dream of jumping from a plane and landing on a net. They worked as a team with a lot of analysis and product testing; under their guidance, he did hundreds of performance tests and thousands of practice sessions (wearing parachutes, of course).

He had such full confidence in the science and engineering behind his goal that when a reporter asked him if people would not think him crazy for taking such a wild risk, he replied '. . . pay attention to the science and the math behind this, and we'll show you what's possible'.

All of it paid off. But let us be realistic.

Notwithstanding the excellent scientific support and his own superior skydiving skills, Aikins was also incredibly lucky everything went by the script all the way. So many things could have gone wrong with phenomena beyond anybody's prediction, let alone control.

Assuming everything else went off perfectly, a single unexpected gust of wind a second before he touched down would have thrown him off the net on to hard ground from 100 ft up, with a tragic ending! That it did not happen was not his doing, but plain luck.

The lesson to be learnt from this episode is yes, in theory, one could jump from any height and still land safely on a chosen spot. But in practice, so many variables are involved which cannot be precisely predicted or maintained, and it would be unwise to risk life and limb for such a (mis)adventure!

Aikins' feat is probably not reproducible, even by himself.

So, we, ordinary mortals, should not tempt fate by exposing ourselves to such extreme risks at the workplace.

All the preceding cases do affirm that safe landing of a fall from height can be planned scientifically, and executed with the proper training and practice.

However, to reiterate a point consistently to be remembered that someone showed it was theoretically possible will not mean that everybody working at height can or should do it!

Many other practical considerations, particularly the capabilities of the actual jumper, will be involved, and a careful evaluation and risk assessment must be conducted before trying anything different!

9.7   BUNGEE JUMPING

Bungee jumping is a voluntary and intentional fall, more a planned and organised fun activity than a task requiring elimination or reduction of injury from accidental falls during a necessary move to and work at height.

The main scientific difference between bungee jumping and other falls is that in this popular 'city/fun' version of bungee, the elastic cord used for the jump allows the person to oscillate up and down a few times before coming to rest and be brought back to the base. In all other cases we have discussed thus far, the jump ends (and is planned to end) on the first contact with the earth or other base.

9.7.1   EVOLUTION OF BUNGEE JUMPING

As described earlier, the idea of tying some elastic rope to the ankles and jumping off from a high perch originated from the millennia-old (and still practiced) tribal custom in the island of Vanuatu off of Australia, mentioned in the very first chapter of this book.

The tribal version used naturally growing vines were more elastic than manufactured ropes but still put a lot of strain on the youth's ankles. Further, they landed with a thud on raked soil at the base of the tower, and hence the jump was not only a test of the bravery and determination of the young men but also the strength and resilience of their bodies.

The business acumen of vacation fun and amusement ride industry folks have converted this to a pleasant and safe excitement for which people pay money, aiming to provide the euphoria of not just jumping from height, but swinging up and down like a yo-yo a few times, soaring like a bird. Many licensed bungee jumping places have become quite popular around the world.

9.7.2 BUNGEE JUMP OPERATIONS AND REGULATIONS

Currently, bungee jumping offered by organisations is a highly regulated activity, permitted only by licensed operators under very strict codes of practice. Hence, it is quite safe, with a global fatality rate of 1 in 500,000.

Many codes ban outright jumping by the ankles and require full-body safety harness systems (without shock absorber), or at least a chest and waist harness and two-point suspension, further limiting the G-factor on the jumper to about 4. Many operators permit ankle-only options with a reduced G-factor limit of around 3. Design factor of safety is a minimum of 5 and often up to 10.

9.7.3 PHYSICS OF BUNGEE JUMPING

Simple as the bungee jump appears (from outside the jump area!), the physics and mathematics of bungee jumps are actually too complex for easy or definitive solution.

There are many solutions based on various assumptions and covering more and more variables, but there is no unanimity on which one is the most appropriate for any situation. Hence the high factor of safety and the low fatality rate.

Figure 9.3(a) displays the various simple harmonic motions of oscillating objects:

(i)   Ideal undamped perpetual sine-wave oscillations;
(ii) Strongly and critically damped motion, which go from maximum d to zero in a finite desired time with no further oscillation, such as soft landing with net or cushion; and,
(iii) Under-damped oscillation of bungee jumping – shown shaded – giving the jumper the thrill of the initial maximum fall, followed by three to five oscillations of decreasing amplitude, and coming to rest before it gets tiresome or boring.

Figure 9.3(b) presents the schematic of a typical bungee jump. The solution for this damped simple harmonic motion involves high-order differential equations.

As far as bungee jump safety goes, we need to be concerned only with the first downward movement, reaching down to the greatest depth from the launch platform, because that would impose the maximum G-force on the jumper.

The elastic cord will extend during the downward fall and store energy, which will be released as kinetic energy in the return upward movement of the body.

Hence the net upward force on the object would be the upward force due to the elastic extension of the bungee cord, minus the weight of the jumper, namely

($k.d - m.g$), where k is the spring constant of bungee cord. Then the G-factor, G, on the jumper is:

$$G = (k.d - m.g)/(m.g).\qquad(9.1)$$

Note that G-factor is not any more $(h+d)/d$, as it is not a free fall.

If we wish to restrict the G-factor to a certain limit, say G_0, then the spring constant k and cord stretch d for a given m are related by:

$$k.d = m.g(G_0 + 1).\qquad(9.2)$$

The main aim with bungee jumping should be to adjust the variables under our control in such a way that the jumper enjoys a challenging and exhilarating experience without suffering too high G-forces.

9.7.4 BUNGEE STRETCH FOR KNOWN FALL

A reasonable approximation to real-life situations, involving the dominant variables, would be to include the bungee-cord weight and its elastic behaviour into the equation. One common (highly simplified) relationship, based on energy balance, whose derivation may be found in Ref. **9.12**, is given here:

$$k.d^2/2 - (m + w/2).g.d - (m + w/4).g.h = 0\qquad(9.3)$$

from which

$$d = \left\{(m + w/2).g + \sqrt{\left[(m + w/2)^2.g^2 + 4(k/2).(m + w/4).g.h\right]}\right\}/k\qquad(9.3a)$$

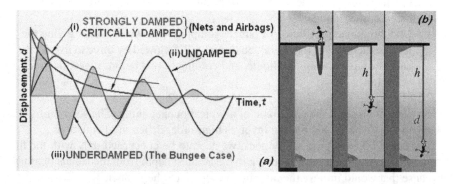

FIGURE 9.3 Bungee jumping theory and practice.

Source and author: Own work, based on classical theory and generic images.
(a) Undamped and damped simple harmonic motion.
(b) Stages in bungee jumping.

$$\text{and, } k = \left(2/d^2\right)\left[(m+w/2).g.d+(m+w/4).g.h\right] \qquad (9.3b)$$

where, in addition to the notation already defined, w is the weight of the bungee cord per unit length; this may be omitted only if it is negligible in comparison with the weight of the jumper.

A major assumption in this formulation is that the spring constant k remains constant over its full stretch d, while in reality, the stiffness decreases as the cord length increases.

Equation (9.3b) gives us a way of accommodating a bungee jump within a certain available total height, say C, because for any given L, the corresponding d, would be $(C–L)$, and then k may be found.

Naturally, in this case, the G-factor should be checked for the cable with the computed k, and confirmed that it is feasible.

9.7.5   EXAMPLE 9.1

A jumper weighing 80 kg jumps with a bungee cord 40 m long and weighing 5 kg. Spring constant of the cord is 100 N/m. What is the stopping distance for the first cycle? What is the G-factor?

Answer:

By Eq. (9.3a), stopping distance,

d = {(80+5/2)×9.81 + $\sqrt{[(80+5/2)^2\times9.81^2+4(100/2)}$ × (80+5/4) × 9.81×40]}/100
= 34.6 m.

For this case, the G-factor is given by Eq. (9.1) as:

GF = (100×34.6–80×9.81)/(80×9.81) = (3460–784.8)/784.8 = 3.41, quite safe.

Note that this implies that a total depth of (40+34.6), that is, 74.6 m plus a safety margin (say) of 5 m, or a total of about 80 m is available for the jump. If this is not confirmed in advance, disaster may result.

9.7.6   EXAMPLE 9.2

If in Example (9.1) the available depth (after allowing for a safety margin) is 75 m, what should be the spring constant? What will be the G-factor?

Answer:

As C = 75 m and L = 40 m, d = 75–40 = 35 m
By Eq. (9.3b), spring constant is given by:

k = (2/35²)[(80+5/2)×9.81 × 35 + (80+5/4)×9.81 × 40)] = 98.3 N/m.

The G-factor for this case is given by Eq. (9.1) as:

GF = (98.3 × 35–80 × 9.81)/(80 × 9.81) = 3.38, quite safe.

9.7.7 CABLE PARAMETERS TO SUIT PRACTICAL CONSIDERATIONS

In Example 9.2, we got a spring constant of 98.3 N/m as appropriate for the situation. But spring constants of manufactured bungee cables come with certain fixed values, so we will have to pick the one closest to our needs from the available choices, in which case the stretch and G-factor for a given length of cord will change as in Example 9.1.

If under these conditions we have constraints on the available jump space, and or maximum G-factors, then we will have to do some more manipulations.

(a)   Given C and G0, to find h and d

Noting that $C = (h+d)$, and using Eq. (9.2) for cable tension $k.d$ (denoted by T), after some rearrangement of terms, Eq. (9.3) reduces to:

$$h = C / \left\{ 1 + (m + w/4).g / \left[T/2 - (m + w/2).g \right] \right\}. \tag{9.4}$$

With h known, d and k are found as $(C–h)$ and (T/d), respectively.

These three theoretical values may now be adjusted by repeated trials, to suit practical considerations, and G checked to be within G_0.

(b) Practical considerations

None of these 'accurate' theoretical values may be workable in practice and may serve only as good guidelines for final decisions.

We may be able to provide an exact length of cable, but the tying process, strap slippages, wind resistance, etc., will affect the final d. Even so, the customer should not be surprised if instead of his fingertip just touching the water as he might have asked for, he misses touching it by a small gap, or his hand goes into the water all the way to the elbow! (Surely the organisers would have got him to sign a waiver on it.)

If the above values d and k are not feasible, a re-analysis with more appropriate values may be suitable to the extent possible.

Eventually, it will all boil down to choosing an available cable with k nearest to what we want, and playing with h and d totalling to C with G less than G_0, and hoping for the best.

9.7.8 EXAMPLE 9.3

The bungee customer wishes to just touch the water which is 75 m below the jump platform, without exceeding a G-factor of 4. If the distance of his ankle strap to fingertip of stretched arm is 2 m, suggest a suitable bungee cord.

Answer:

Required $C = 75-2 = 73$ m.

We will assume that the cord weight will still be 5 kg.

For $G_0 = 4$, $T = k.d = 80 \times 9.81(4+1) = 3924$

For these values, Eq. (9.4) gives:

$$h = 73 / \{1+(80+5/4)\times9.81/[3924/2-(80+5/2).9.81]\} = 43.16 \text{ m.}$$

Then, $d = 73-43.16 = \underline{29.84 \text{ m}}$

$$k = 3924/29.84 = \underline{131.5 \text{ N/m.}}$$

With these theoretical values as guidelines, we may propose a practical solution.

For instance, if we set the cable k at 125 as the nearest available, we can arrive at the combination of h and d to satisfy the governing Eq. (9.4). Each time d changes, h obtained as (73-d) changes; we update h after each cycle to get the new d and repeat the process until sufficient convergence is obtained, retaining the k at 125.

Table 9.1 shows the iterative process, which in five cycles converged to two-decimal accuracy.

With $k.d = (125 \times 30.61) = 3826$, the G-factor for the final choice of 42.39 and 30.61 m for h and d becomes, by Eq. (9.1):

$$GF = (3826-785)/785, \text{ that is, } \underline{3.87}, \text{ confirming less than limit of 4.}$$

9.7.9 BUNGEE APPLICATION TO FALL CONTROL

The primary justification for using up all this space on a topic which may not have much application to real-life fall management is that it illustrates how we can control our own falls to a great extent. A secondarily reason is that it has had some real-life applications.

(a) James Bond (007) 'Golden Eye' movie, 1995

One of the first bungee jumps, long before its physics had been well analysed, was in the 1995 James Bond movie 'Golden Eye' wherein the hero Peter Brosnan – actually

---

TABLE 9.1

Iterations to determine d for given C and G.

Iteration	1	2	3	4	5	Remarks
h	43.16	42.19	42.44	42.38	42.39	$h_n = 73-d_{n-1}$
d	30.81	30.56	30.62	30.61	30.61	d from Eq. (9.4)

his stunt double Wayne Michaels – bungee jumps down the 220 m Verzasca Dam in Switzerland.

With milliseconds to spare at the bottom of his first drop, 007 pulls out a dart gun and fires into his chosen landing spot a piton attached to a rope and then reels himself to safety. He did not, of course, oscillate! – that would not have been 'Bond'-ish!

It held the record for the longest jump from a fixed object for a long time and also bagged the award for the best movie stunt.

Since then, many movies around the world have incorporated a bungee jumping scene, in some cases by the actors themselves – like the South Indian movie star Vijay did in 2000 for the movie 'Kushi'.

(b)   Considered in Luke Aikins' 25,000 ft jump

Bungee jump figured seriously also in the planning of Aikin's 25,000 feet jump from the plane!

Civil engineer John Cruikshank who designed Luke Aikins' fall arrest system did consider bungee cords for deceleration before he decided on or air pistons [Ref. **9.13**].

The problem with bungee cords was that while they would slow Aikins down, he would be thrown back up in the air after the first drop, and subsequent events might go awry or chaotic!

Of course, for this book, discussion of bungee jump is only of academic interest. Hopefully it has given a glimpse of what is involved in soft landing, and what is, at least in theory, possible. I hope the readers enjoyed it as much as I enjoyed writing it!

9.8   RISK ASSESSMENT FOR WORK AT HEIGHT

Risk assessment is an essential prerequisite and a powerful predictive tool for mishap prevention and control. That is why, before any work at height is undertaken, many countries insist on a risk assessment exercise for falls as well as injuries from them.

Most commonly, risk is assessed as the combination of:

(a) Likelihood of occurrence of a hazard;
(b) Severity of consequences of the mishap if it should occur.

On the face of it, fall prevention methods (of Chapters 3 and 4) will eliminate the need for further risk assessment of risk of falls. However, in real-life statistics, there is no 0% and no 100%; even with fall prevention, the likelihood of fall will not be zero, but low or very low if you wish. For example, a flimsy or poorly fixed guardrail may give way if a worker bumps into it.

9.8.1   LIKELIHOOD OF FALL

In the case of accessing or working at height, the overall likelihood of fall injuries may be estimated as follows:

- Low, if fall prevention methods of Chapters 3 and 4 are fully implemented;
- Medium, if the fall arrest methods of Chapters 6–8 are fully implemented;
- High, if the recommended safeguards are not fully implemented.

Understandably, the actual likelihood level will depend on the specific circumstances governing any particular case.

9.8.2 Severity of fall injury

The severity of fall injury will depend on several factors, all of which are not amenable to close prediction or evaluation. The range of fall injuries also will be very wide as has been explained in previous chapters.

In general, the following may be postulated:

- Low, if the fall is on external or wearable airbag;
- Medium, if the fall is on safety nets or with safety harness and shock absorber;
- High, if the fall is on hard surface and without any PPE.

Again, the actual severity level will depend on the specific circumstances governing the particular case.

9.8.3 Risk of fall injury

Conventionally, risk category is assessed from the combination of likelihood and severity levels, typically as follows:

- Low, if both component levels are Low, or if one is Low and other is Medium;
- Medium, if both components are Medium, or if one is Low and other is High;
- High, if both components are High, or if one is Medium and other is High.

Once the risk category is assessed, the controls may be recommended along the lines already discussed, adopting the overall guidelines as follows:

- Low Risk: 'Acceptable'; no need for further safeguards, except monitoring against deterioration;
- Medium Risk: 'Manageable'; control with appropriate safeguards and continuous supervision;
- High Risk: 'Unacceptable'; do not start, stop if started, until risk is brought down to at least Medium.

Most of the time, even with the best of controls, the pragmatic and conservative approach would be to take both likelihood and severity as medium, leading to medium risk. Thus, continuous and careful attention will be called for, during any and all work at height.

Further information on risk management may be gathered from my book on it, Ref. **9.14**.

9.9 EXERCISES FOR CHAPTER 9

9.9.1 EXERCISE 9.1

A jumper weighing 90 kg uses a bungee cord 50 m long, with spring constant of 150 N/m and weighing 0.22 kg/m. What are the stretch and the G-factor?

Answer:

$d = 31.65$ m; $G = 4.4$.

9.9.2 EXERCISE 9.2

What might have been the spring constant for the bungee cord that James Bond used for the 220 m jump, if the elongation was 50% and he had to shoot a piton when he was 10 m above the base to reel himself in? Assume Bond's weight at 80 kg and cord weight at 20 kg. What maximum G-force would he have felt?

Answer:

$k = 72.87$ N/m; $F = 4.3$ kN.

9.9.3 EXERCISE 9.3

Let us say that Aikins' planners had chosen to stop him with a bungee system of four cords. It was determined that he would reach the landing site with a velocity of 190 km/h. The planners used a weight of 90 kg for the design and expected his G-factor to be not more than 5. Assume that each of the four cords shares the impact equally. What combination of k and d would (theoretically) have worked if the cord elongations had to be kept around 50%?

(<u>Hint</u>: Get height of fall equivalent to the velocity, which should be in m/s. Take $d = h/2$ and $G_0 = 5$, and get k from Eq. 9.2.)

Answer:

<u>$d = 71$ m</u>; $k = 74.61$ N/m divided into 4 cords, or <u>18.65 N/m</u> each.

[It just happens in this case, any variation from these values would violate one or the other of the requirements! It can be confirmed that increasing k will increase G. If the feasible k lower than 74.61 is <u>72 N/m</u>, d will become <u>72.54 m</u>, elongation will be <u>51.1%</u>, marginally > 50%, and G will be <u>4.92</u>, < 5.0.]

REFERENCES FOR CHAPTER 9

9.1 ———. *Fact Sheet on Fall Protection Plan*. Ministry of Manpower and Workplace Safety and Health Council, Singapore, 021209. Retrieved on 13 March 2023 from: www.mom.gov.sg/-/media/mom/documents/press-releases/2009/annex-c-fact-sheet-on-fall-protection-plan-(021209).pdf

9.2   ———. *Model Fall Protection Plan*. OSHA, USA. Retrieved on 15 August 2023 from: www.osha.gov/sites/default/files/2020-03/Model%20Fall%20Protection%20 Plan.pdf

9.3   ———. *Guidelines for Creation of a Fall Rescue Plan*. Associated General Contractors of America, NSC Construction & Utilities Division, National Safety Council, USA. Retrieved on 15 August 2023 from: www.nsc.org/getmedia/e459a825-5d2b-4244-bfae-857d147a5d67/planning-fall-rescue-plan.pdf.aspx

9.4   Knight, N., "WW2 airman fell 20,000 feet without a parachute before smashing through the glass roof of a train station. He survived and lived to 82", *Vintage News*, 17 April 2016. Retrieved on 15 March 2023 from: www.thevintagenews. com/2016/04/17/ww2-airman-fell-20000-feet-without-a-parachute-before-smashing-through-the-glass-roof-of-a-train-station-he-survived-and-lived-to-82/?chrome=1

9.5   Wills, K., Deutsch, K., and Hutchison, B., "Miracle! Man plunges from W. Side building, crashes atop of Dodge Charger and lives to tell tale", *New York Daily News*, 1 September 2010. Retrieved on 17 March 2023 from: www.nydailynews.com/new-york/miracle-man-plunges-w-side-building-crashes-atop-dodge-charger-lives-tale-article-1.202576

9.6   Reuters Staff, "Argentine woman survives fall from 23rd floor", *Reuters – Lifestyle*, 25 January 2011. Retrieved on 17 March 2023 from: www.reuters.com/article/ us-argentina-fall-idUSTRE70N6OR20110124

9.7   ———. *World's First Ever Wingsuit Landing Without a Parachute*. Retrieved on 16 March 2023 from: www.youtube.com/watch?v=DEP8juRSBRo

9.8   ———. "Gary Connery", *Wikipedia*. Retrieved on 16 March 2023 from: https:// en.wikipedia.org/wiki/Gary_Connery

9.9   ———. "Worldrecord: 45 meters free fall into boxes – YouTube", *Slamartist.com*. Retrieved on 15 Mar. 2023 from: www.youtube.com/watch?v=jRbqpctO4rQ

9.10 ———. "David Blaine", *Wikipedia*. Retrieved on 12 July 2023 from: https://en.wiki-pedia.org/wiki/David_Blaine

9.11 ———. "The extreme engineering that made Luke Aikins' historic skydive possible", *Create Digital*, 19 July 2018. Retrieved on 18 March 2023, from: https://createdigital. org.au/extreme-engineering-luke-aikins-skydive/

9.12 ———. "The physics of bungee jumping", *Real World Physics Problems*. Retrieved on 21 March 2023 from: www.real-world-physics-problems.com/physics-of-bungee-jumping.html

9.13 ———. "The extreme engineering that made Luke Aikins' historic skydive possible", *Create Digital*, 19 July 2018. Retrieved on 18 March 2023 from: https://createdigital. org.au/extreme-engineering-luke-aikins-skydive/

9.14 Krishnamurthy, N., *Introduction to Enterprise Risk Management*. Partridge Publishing, USA, 262 pp, 2019. ISBN-13: 978-1543754742.

10 Other fall-related topics

10.1 CONSIDERATIONS OTHER THAN FALLS FROM HEIGHT

There are a few topics not directly related to human falls from height during routine daily life or work, but which involve people falling at level, or objects falling with potential hazards for human harm, and requiring design and controls to eliminate the mishaps or mitigate their consequences.

In particular, the following will be relevant:

- Slips, trips, and falls
- Falling objects hazards
- Recent developments in fall management.

These will be dealt with in the following sections.

10.2 SLIPS, TRIPS, AND FALLS

Although we had included falls at level with falls from height in our earlier discussions, falling at level has one important difference from falling from height.

That is, most of the analysis and conclusions therefrom, which were presented in the earlier chapters, will not apply to the falling from level cases.

Slips, trips, and falls are often the largest contributor to non-fatal fall accidents.

Many times, the person who slips or trips is quick, smart, or capable enough to recover and avoid a fall.

But all too often, even falling from level may involve a sudden crashing of the upper regions of the person's body on to the ground or other base with great force, causing him grievous harm or even death.

10.2.1 DEFINITIONS

Figure 10.1 illustrates slips, trips, and falls, abbreviated to STF.

(a) Slips

Slips are when the person's foot slides on the floor losing its grip and allowing, even causing, the person to fall, usually backwards, and collapse to the floor, unless a fall is saved by the person grabbing some support or someone holding the person up.

This may happen when the person steps on some lubricating fluid, small granular masses, or loose coverings on the floor. It may be aggravated or even initiated by very smooth shoe soles unable to develop the necessary friction for the person to move forward.

DOI: 10.1201/9781032648132-10

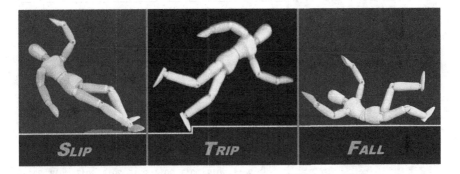

FIGURE 10.1   Slips, trips, and falls (STF).

Source and Author: Own work.

(b)   Trips

Trips are when the person's foot hits a small stationary or movable obstacle on the floor but his body keeps moving forward and he stumbles to the floor, crashing, usually forward, to the ground or floor, unless again, the person grabs a support or is grabbed by someone.

The obstacle may be a cable or rope, wrinkled carpet, or any small hard object or debris, or even sloping or uneven floor surfaces.

Poor lighting or housekeeping can aggravate the slip and trip hazards. Both slip and trip hazards usually end up in sudden falls at level, with minor injuries if the collapse is on the back or buttocks or forwards with support from extended arms, but resulting in major injuries or death if the back of head or face hits any hard surface.

(c)   Falls

Falls – here specifically referring to falls at level – are when a person collapses on a level or slightly sloping surface without slipping or tripping, as may happen when he has a sudden spell of vertigo or dizziness, or he has been standing still in place for too long.

Sudden stoppage of movement or sudden jerk forward also may lead to a fall.

As falls from height are usually defined as those happening above 2 m, short falls as from the lower ladder rungs or stair steps are also included in this category.

10.2.2   STATISTICS OF SLIPS, TRIPS, AND FALLS

Figure 10.2(a,b) displays statistics from two sources, one from New Brunswick, Canada and the other from the Bureau of Labor Statistics, on slips, trips, and falls at same level, together with falls from height.

Without going into details, it may be arguably claimed that STF constitute a significant part of fall injuries. Other sources [Ref. **10.1**] indicate that STF fatalities are no small part of the total number of deaths at home or workplace.

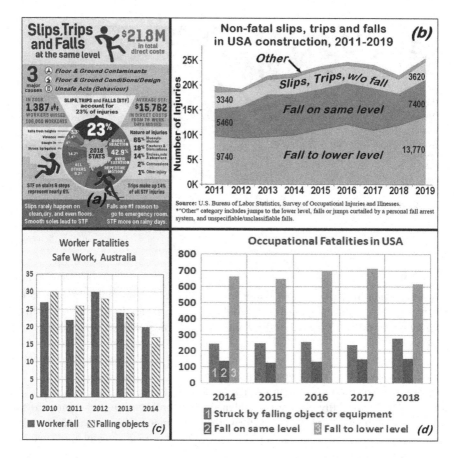

FIGURE 10.2   Statistics of person slips, trips, and falls, and falling objects.

(a) STF in Canada. [Source and Author: Own work, redrawn from Infograph of Worksafe, New Brunswick, Canada; Link: www.worksafenb.ca/safety-topics/slips-trips-and-falls/]

(b) Non-fatal slips, trips, and falls in the USA Construction. [Source: Survey of occupational injuries and diseases. Author: Bureau of Labor Statistics. Link: https://blogs.cdc.gov/niosh-science-blog/2021/04/05/falls-standdown-2021/]

(c) Worker fatalities in Australia. [Source: Safe Work, Australia. Author: Own Work, plotted from source data in Ref. **10.2**. Link: www.linkedin.com/pulse/dropped-objects-really-second-highest-cause-workplace-rick-salisbury]

(d) Construction fatalities in the USA. [Source: News Release, Bureau of Labor Statistics, USA, 17 Dec. 2019. Author: Own work, plotted from source data. Link: www.bls.gov/news.release/archives/cfoi_12172019.pdf

The situation is very much the same globally. Overall, STF may be second to falls from height in fatalities, but is often the first in major injuries.

This should ring a warning bell on the importance of the usually not seriously taken slips, trips, and falls at level.

10.2.3 CONTROL OF SLIPS, TRIPS, AND FALLS

While the avoidance or management of STF may appear basic and simple in the sense that we intuitively do that all the time we move around, facts speak to the high prevalence of STF accidents at the workplace, costing huge amounts of money in many countries.

Common control measures for the management of slips, trips, and low falls by various stakeholders at the workplace may be summarised as follows [Ref. **10.3**].

(a) Employers – planning and workplace design

- Develop a written STF prevention policy that specifies both employer and worker responsibilities.
- Ensure that aisles and passageways are free of clutter and other tripping hazards.
- Provide proper lighting in all areas to reduce shadows, dark areas, and glare. Replace burnt out light bulbs promptly.
- If electrical cords are used on a regular basis, install outlets so that cords do not cross walkways.
- Ensure that water or other liquid sprays done for required procedures are not sprayed onto the floor.
- Provide clean-up supplies (paper towels, absorbent material, "wet floor" signs, etc.) at convenient locations in the site.
- Select flooring material according to the work to be done in the area. Use flooring with a static coefficient of friction of more than 0.5 for high-risk areas.
- Insist on keeping floor surfaces clean and dry.
- Check and eliminate obstructions to walkers' vision.
- Provide water-absorbent mats near entrances and other areas where water, ice, or snow may drip or be tracked onto the floor.

(b) Employers – training

- Train employees to identify STF hazards and how to prevent STFs by using safe cleaning procedures, including placing cones, triangles, and/or caution signs around the site.
- Make sure employees know whom to call to report hazards and whom to call for clean-up or repair.

(c) Employers – footwear

- Insist on slip-resistant shoes for staff that work on wet or contaminated walking surfaces.
- Require footwear that is resistant to oil, chemicals, and heat.

(d) Employees – material handling

- Walk with caution and make wide turns at corners.
- Push (rather than pull) carts to allow a better line of sight.

(e) Employees – housekeeping
- Clean floors and work surfaces as soon as they become wet.
- Place warning signs in wet floor areas and remove them promptly when the floor is clean and dry.
- During wet or oily processes, maintain drainage and provide false floors, platforms, or nonslip mats.
- Clean only one side of a passageway at a time to allow room for walkers passing.
- Keep passageways clear at all times, and mark permanent aisles and passageways.
- For purposes of infrequent use, tape or anchor electrical cords or other cables to floors if they cross walkways.
- Clean off any slippery material on the rungs, steps, or feet of a ladder before you use it.

10.3   FALLING OBJECT HAZARDS

10.3.1   THE PROBLEM WITH FALLING OBJECTS

At work, only second to fatalities and injuries due to human falls are the fatalities and injuries due to objects falling on humans. Typical of the magnitude of this world-wide problem is evident from Figure 10.2(c), displaying the relevant statistics from Australia, and Figure 10.2(d), similar ones from the USA.

The Australian data indicating nearly equal fatality numbers for falling from height and for objects falling on persons is unusual. A more common finding globally is more like the object fall fatalities being about a third of fall from height fatalities.

What objects fall? – Tools and materials being used by the workers at height, debris broken off from the work at height, broken or collapsed portions of platform or support structure for the workers etc.

As has been discussed at length in Chapter 5, every object falling from height picks up velocity. As most falling objects are of dense materials and have sharp edges, any object happening to hit a human can definitely have highly serious conse-quences including death.

What is more disconcerting is that falling objects may not just fall straight down. Data has shown that a tool weighing 3.7 kg falling from a 22-storey building and hitting a bar 6 m off the ground can deflect up to 127 m, reaching a top speed of over 130 km/h!

Debris drop zones and traditional barricading methods are largely rendered inef-fective in this type of scenario.

10.3.2   HIERARCHY OF FALLING OBJECT CONTROL

As with humans, the hierarchy of falling object control should follow a logical sequence similar to that used for human falls, as follows:

(a)   Collective object fall prevention
Toeboards on all working platforms; warning signs; barricading around fall area; debris screens around the structure under construction, Figure 10.3(a); debris catch-nets projecting out from the building faces, Figure 10.3(b), more details of which have been discussed in Section 6.3; and effective house-keeping.

FIGURE 10.3 Controls for falling objects.

(a) Debris net around building. [Source and Author: Freepik. Link: www.freepik.com/pre-mium-photo/building-facade-renovation-works-with-construction-scaffolding-frame-covered-with-protective-net-workers-safety_27904398.htm]

(b) Debris catch-net. [Source and Author: Karam Safety Private Limited, India. Screen-shot from video, with permission from the author. Link: www.youtube.com/watch?v=-jQy-om8bVyM]

(c) Fall protection canopy. [Source and Author: Own Work, background image from public domain.]

(d) Tethered tools. [Source: Unsplash. Author: Photo by Dean Bennett. Link: https://unsplash.com/photos/aBV8pVODWiM]

(b) Individual object fall prevention

Tethers to all tools used by every worker [see Figure 10.3(c)]. Objects for tethering either have built-in connection points placed by the manufacturer or can be retrofitted with connection points, at which they are connected to a lanyard, with their other end attached to the waist belt of the worker. Tethered tools are not in wide use yet. But this may change, as younger workers become more aware and concerned with safety than veterans.

(c)   Collective fall arrest of objects

Canopies, or 'Fans', as temporary shelters for walkways adjacent to construction sites [see Figure 10.3(d)] to protect pedestrians from falling objects. The shelter roofing sheets usually are laid with a slight downward slope towards the construction side so that if any debris falls on the sheet, it will bounce into the site instead of the road or other public space outside the site.

(d)   Individual fall arrest of objects

Hard hats to all personnel, so commonly used all over the world.

Perhaps not surprisingly, the last one in the list, namely hard hats are the most commonly used control at work sites. Unlike the safety harness, hard hats are easy to wear and are quite effective in protecting the head from injury for most small tools or debris falling from low heights, but completely useless for heavier objects and larger heights, demanding other safety measures.

Recently, the traditional hard hat has undergone a drastic change deserving the name 'safety helmet', which has thicker cushions and suspension straps, which transfer and soften the impact of falling objects on it.

10.3.3   CHALLENGES FACED BY INDUSTRY

The challenges to industry posed by these requirements are very real and tend to make it difficult for tool drop prevention polices and procedure to be implemented [Ref. **10.2**]. Some common concerns and reasons for hesitancy are as follows.

(a)   Poor safety culture

Tool drop prevention is overlooked in many industries and is possibly not as easy to comprehend as some of the more recognised industrial hazards.

(b)   Lack of risk awareness

Familiarity with the risk of falling objects on work sites may have tended to make us complacency, despite evidence that this may have deadly consequences.

(c)   Restricted tool functionality

Historically, many tool-drop prevention devices have inhibited tool functionality and, therefore, affected job performance, discouraging workers from using them.

(d)   Tool diversity

The vast range of tools used at height presents a very challenging situation for a drop prevention policy to be implemented, preventing a 'one size fits all' approach.

10.3.4   DROPPED OBJECTS BEST PRACTICES AND SOLUTIONS

Top best practices and solutions for organisations when it comes to dropped object prevention [Ref. **10.4**] are as follows:

(1) Expand fall protection programs to include tools and equipment.

(2) Provide a competent person to manage the expanded program.

(3) Raise awareness of drop hazard identification and mitigation techniques within the workforce.

(4) Require risk assessments before performing work with drop hazards.

(5) Consider regularly scheduled 'hazard hunts' to drive awareness of drop hazards.

(6) Consider using tethered tools and small equipment items.

(7) Consider using energy-absorbing lanyards, which will reduce the force associated with the dropped tool. Tools can be either connected to a worker through a tool belt, harness or wristband, or anchored to a fixed structure.

(8) When a worker needs to pass his tool off to a colleague, the colleague can connect to the tool before the passing worker disconnects from it, to ensure 100% tie-off.

(9) Workers at height should bring up only the tools they need to do their job.

(10) Hoist up items and then transfer them over with different lanyards to the workers themselves or to static anchor points. This can be done in a bucket, which can then house the extra tools, but spills during transfer must be avoided.

(11) Where work platforms are used, install toeboards, capable of withstanding a force of at least 20 kg (44 lb) in any direction.

(12) Debris nets provide a way to catch dropped objects. Netting covering the façade of buildings is the most common, but the netting can also be put up within the construction, such as directly under work being done. The mesh should be small enough to catch the smallest object likely to fall.

10.3.5 REQUIREMENTS FOR TOOLS

A 'best practice' Tool Drop Prevention System that meets the following requirements [Ref. **10.2**]:

- Is innately adaptable to most of the hand tools used;
- Takes a short time to apply and does not require technical expertise;
- Does not require tools to be mechanically modified, or get damaged in any way;
- Provides a fully certified and load-rated system for industry;
- Provides a viable avenue for companies to become aligned with Drop Object Prevention legislation.

10.4 RECENT DEVELOPMENTS IN FALL MANAGEMENT

While in the past, innovations used to take decades to develop and be implemented, we now have new products almost every week. It is the same with work at height safety.

In the last few decades, many technological advances have taken place, which not only address the new work-at-height hazards that appear with every new development in construction and manufacturing technology but also raised the detection of existing hazards and their management to new levels of efficiency.

Among them may be cited the following.

10.4.1 RADIO FREQUENCY IDENTIFICATION

Radio frequency identifications (RFIDs) have been in use for decades as tracking and locating devices. They consist of radio-wave transmitters ('tags') and receivers ('readers'), which can be attached to objects or worn by persons as the situation demands. The tags broadcast location and other information, which the reader(s) analyses and acts upon as programmed by software.

While this has found multiple uses in inventory control, equipment and worker tracking etc. specifically, it is being used to check and warn a person nearing an unprotected edge or other fall risk, as schematically indicated in Figure 10.4(a). In the figure, the worker carrying a ladder nears an unprotected edge transmits radio signals from the tag (T) he wears. Readers (R) suitably positioned around the safe zone keep analysing location information on the tag as he moves around, with the software in the control centre (C).

If he nears the unsafe zone (such as the unprotected edge at height as shown here) to a prearranged distance, say 2 m, it will give him a warning buzz; if he goes further to a more critical distance, say 1 m, it will sound a loud siren or by other means either automatically or through the control technician, stop him from further exposure to the risk.

RFID has certain limitations like signal collision between simultaneous multiple signals and susceptibility to data theft and manipulation. But with improving technology and in combination with other developing technologies, its versatility and applicability are increasing.

10.4.2 BUILDING INFORMATION MANAGEMENT

Building information management (BIM) has now been in use for decades, effectively enabling the user to examine from various angles – in effect, to walk through – a building project even before construction has started. From this, a safety manager may:

(a) Identify zones with fall at height hazards;
(b) Check if all potential fall risks have been addressed in the permanent picture;
(c) Visualise fall hazard management such as safe zones for fall restraint, and recommend fall controls such as edge protection and anchor locations, during the construction or erection phase;
(d) Utilise the BIM model as the virtual world for worker tracking, virtual reality training etc.

Reference 10.5 gives a broad description of how BIM may be used for various hazard identification and control purposes. Figure 10.4(b) from the paper displays a typical

FIGURE 10.4   Recent technologies in fall management – I.

(a) RFID. [Source: Source and Author: Own work, using worker clipart from Freepik.]

(b) BIM for fall control [Source: Conference Proceedings Ref. **10.5**. Authors: Azhar, Behringer et al. as in Ref. **10.5**. Image used with permission from the authors.]

(c) Conventional bridge inspection. [Source: MDPI Open Access Article Ref. **10.6**, Figure 1(b). Authors: Cano, Pastor et al of Ref. **10.6**. Link: www.mdpi.com/2072-4292/14/5/1244]

(d) Bridge inspection by drone. [Source and Author: *ibid*, Graphical Abstract.]

(e) Drone passes for inspection. [Source and Author: *ibid*, Graphical Abstract.]

application. It may be seen that both edges of the sloping roof panel AB, as well as the beam edge D, are potential fall hazards, requiring planning and fall control by edge protection (guardrail) or soft landing (safety net) to eliminate or mitigate consequences of fall risk.

10.4.3   Geographical information system

Geographical information system (GIS), which displays topographical maps of an area with overlays of man-made surface and underground features, will be quite useful to identify potential problems during excavation stages for any construction or extraction, where fall into excavations will be likely, and to design suitable preventive and arrest measures.

10.4.4 GLOBAL POSITIONING SYSTEM

Global positioning system (GPS), a navigation satellite system that gives geolocation and time information of any object on earth visible to four or more of the many satellites orbiting the earth, will give macro views of structures that are in place to enable:

(a) Visualisation of support and anchor locations for work-at-height safety measures;

(b) Planning of access by emergency services, rescue measures etc.

10.4.5 DRONES

Drones, which are small unmanned aerial vehicles (UAVs) equipped with cameras and other sensors, and controlled by a relatively stationary pilot on firm ground or other base, are a blessing for the management of much work at height because, without sending persons to physically access the height or depth and carry out the hazardous task, they can:

(a) Transmit to the safely stationed drone pilot and to the data collection and analysis system, visual data – and even thermal, wind and other data as instrumented – from risky-to-reach locations in any structure, already built or under construction, for example, high ceilings, inside utility areas, plan views, under bridges, inside confined spaces etc.;

(b) Deliver small tools and equipment, drawings, safe work procedures, essential supplies etc. to personnel already at the risky spot.

Figure 10.4(c) from Ref. **10.6** depicts the conventional method used for decades to inspect the sides and soffit of a bridge from an articulated boom hoist on a truck. This was cumbersome, time-consuming, and equipment-intensive and held its own risks to the occupants of the boom basket; occasionally, the truck itself overturned and caused more loss!

Now, after the advent of drones, the expense, effort, and risk to inspectors have been completely eliminated, as further described in Ref. **10.6**. Figure 10.4(d) depicts inspection of a bridge by a drone. An experienced drone pilot can chalk out flight paths for drones, as marked in Figure 10.4(e), to efficiently collect all the needed data in a fraction of the time that any other method would take. He can even change or expand his coverage as his screen continuously displays what the drone camera is seeing.

10.4.6 EXO-SKELETONS

Tasks that involve repeated and fatiguing actions by workers may be facilitated by mechanical and electronic augmentation of physical capacity and resilience through devices attached to limbs or enclosing the entire human body. These are called 'exoskeletons'.

Exoskeletons are now commonly used to retain and increase the strength of aging personnel in lifting weights, postures that involve poor ergonomics such as bending, squatting, and reaching up.

Figure 10.5(a) displays an oldish person lifting a box up to or down from the shelf, obviously aided by the exoskeleton around his shoulders and arms.

Countries like Japan which have a large labour shortage but adequate aging population are already able to carry on duties requiring heavy effort by equipping their older workers with exoskeletons, which can not only make up for their enfeebled strength but also facilitate their doing more than their younger colleagues with their greater experience, by smartly manipulating their exoskeletons outfitted with magnified capabilities!

10.4.7 INTERNET OF THINGS

The Internet of things (IoT) refers basically to the data collection, analysis, and control implementation of an ongoing process, facilitated by the Internet (or organisation-wide Intranet) and appropriate software of interactions between man and machine, by means of sensors worn or devices accessed by persons, with the aim of identifying poor posture and other health problems of site personnel so that safety interventions may be made before mishaps occur. Figure 10.5(b) depicts the scope of IoT, which may be seen to (naturally!) include activities involving heights.

Wearables are available, as shown in Figure 10.5(c), to warn owners regarding personal health symptoms and consequences arising from the risks they were taking in the conduct of any hazardous work such as working at heights.

Thus, wearables may be effectively used to warn the wearer working at heights not only of his vital signs but also to communicate essential fall hazard data to the service centre for further analysis and action plans; for example, detection and transmission of high or low blood pressure, temperature, heart rate, breathing difficulties, and STF.

10.4.8 VIRTUAL REALITY

This relatively new immersive technology (abbreviated as 'VR') transports a user with VR headsets and gloves into a virtual world, which can be programmed by computer software to realistically present to the user a 3-D imaginary world with all the hazards of the real world but without the user actually undergoing any of the physical risk or its consequences.

A person can thus be made to face high fall-risk situations such as erecting scaffolds and fixing harness anchors even while the user, although stationary and safe in a studio, soon is convinced that he is actually in the pictured zone and facing the hazard. It is being commonly used for driver and pilot training, with its ability to pose real-life complexities and difficulties without the trainer or trainee facing any actual risk.

Obviously, it is a boon for work at height training. Figure 10.5(d) depicts a trainee (at the bottom right corner) with headset (H) and control gloves (G) with which he enters a construction site (standing in place or moving a little on a firm floor, let us not forget!). He can be presented with a collection (A) of PPE, from which he must select the appropriate ones – and when their images are touched, they will automatically attach themselves to his image viewed by him.

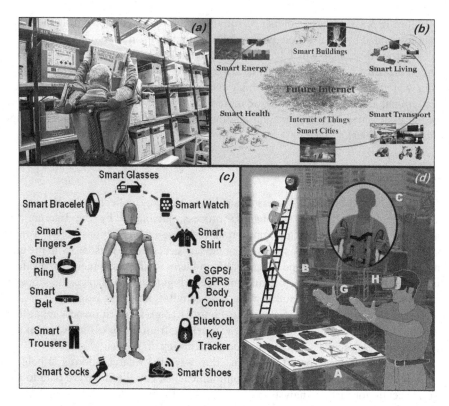

FIGURE 10.5   Recent technologies in fall management – II.

(a) Exoskeleton, to lift heavy box. [Source: Own work. Author: UsaChan93. Link: https://commons.wikimedia.org/wiki/File:MATE-XT_ESOSCHELETRO.jpg]
(b) Scope of Internet of Things. [Source: Own work. Author: Ameer Nasrallah. Link: https://commons.wikimedia.org/wiki/File:Iot_apps.png]
(c) IoT wearables. [Source and Author: Own Work, background figures from public domain.]
(d) Virtual reality for safety training. [Source and Author: Own work, utilising generic images from Freepik.]

He can be trained and tested in the proper use of SRLs for ladders (B) and of double lanyards for 100% tie-off (C).

There are arguably more innovations in the offing. Our concern should still be whether the novice understands the uses and the limitations of the modern techniques or whether he trusts them blindly without ensuring their proper application and contributes his share of the safety management to the process.

REFERENCES FOR CHAPTER 10

10.1 Cayless, S.M., "Slip, trip and fall accidents: Relationship to building features and use of coroners' reports in ascribing cause", *Applied Ergonomics*, 32(2): pp. 155–162, April 2001. Retrieved on 16 August 2023 from: www.sciencedirect.com/science/article/abs/pii/S0003687000000521

10.2 Salisbury, R., "Are 'dropped objects' really the second highest cause of fatalities in the workplace?", *techniquesolutions.com.au*, GRIPPS Global Pty Ltd, USA, 24 May 2016. Retrieved on 24 March 2023 from: www.linkedin.com/pulse/dropped-objects-really-second-highest-cause-workplace-rick-salisbury

10.3 ———. "Preventing slips, trips, and falls in wholesale and retail trade establishments", *Workplace Solutions*, National Institute for Occupational Safety and Health (NIOSH), USA. Retrieved on 27 March 2023 from: www.cdc.gov/niosh/docs/2013-100/pdfs/2013-100.pdf

10.4 Caldwell, M., "The sky isn't falling (and your tools shouldn't either)", *EHS Today*, 1 March 2016. Retrieved on 24 March 2023 from: www.ehstoday.com/construction/article/21917546/the-sky-isnt-falling-and-your-tools-shouldnt-either

10.5 Azhar, S., Behringer, A., Sattineni, A., and Maqsood, T., "BIM for facilitating construction safety planning and management at jobsites", *Proceedings of the CIB-W099 International Conference: Modelling and Building Safety*, Singapore, September 2012.

10.6 Cano, M., Pastor, J.L., Tomás, R., and Riquelme, A., "A new methodology for bridge inspections in linear infrastructures from optical images and HD videos obtained by UAV", MDPI Open Access Journal, *Remote Sensing*, 14(5), 2022. https://doi.org/10.3390/rs14051244.

11 Conclusion

In this book, I have covered many aspects of reaching and returning from heights, for working or having fun at heights that might interest most engineers and hopefully many non-engineers – because falling from height is not just a workplace and occupational safety and health concern, but also a common situation that all of us, of all ages from all walks of life and from all over the world, have been facing since the beginning of human life on our earth.

I have been careful to avoid getting into details of what different countries and different industries in any country have been and are doing to analyse and control injuries from falls, except while mentioning some code or citing some statistic solely to illustrate current thinking on the subject and clarify or support a point I was making and thus enlarge the reader's understanding of a complicated issue.

I have dwelt in some detail on a few esoteric subjects such as horizontal lifelines and bungee jumping, because they are complex and most people avoid it thinking that it is above their understanding.

My intention in choosing such forays into a few complex analyses was to bring home the basic logic and some description of the scope of the many facets of the dangerous activity at height, and remove some of the mystique behind them, and not that anybody could start solving such complicated problems. Frankly, they are indeed complicated in analysis, design, and implementation and should be left to the experts.

That said, if readers have grasped the seriousness of falls even from 'low' heights and believe that their consequences are neither as complex as 'rocket science' nor as arbitrary as authoritarian whim, my purpose would have been served.

Go on, soar to heights . . . but with care!

DOI: 10.1201/9781032648132-11

Index

Printed in the United States
by Baker & Taylor Publisher Services